Student Edition

Eureka Math
Grade 8
Modules 6 and 7

Special thanks go to the Gordan A. Cain Center and to the Department of Mathematics at Louisiana State University for their support in the development of *Eureka Math*.

Published by the non-profit Great Minds

Printed in the U.S.A.
This book may be purchased from the publisher at eureka-math.org
10 9 8 7 6 5 4 3 2 1

ISBN 978-1-63255-322-5

Lesson 1: Modeling Linear Relationships

Classwork

Example 1: Logging On

Lenore has just purchased a tablet computer, and she is considering purchasing an Internet access plan so that she can connect to the Internet wirelessly from virtually anywhere in the world. One company offers an Internet access plan so that when a person connects to the company's wireless network, the person is charged a fixed access fee for connecting *plus* an amount for the number of minutes connected based upon a constant usage rate in dollars per minute.

Lenore is considering this company's plan, but the company's advertisement does not state how much the fixed access fee for connecting is, nor does it state the usage rate. However, the company's website says that a 10-minute session costs $\$0.40$, a 20-minute session costs $\$0.70$, and a 30-minute session costs $\$1.00$. Lenore decides to use these pieces of information to determine both the fixed access fee for connecting and the usage rate.

Exercises 1–6

1. Lenore makes a table of this information and a graph where number of minutes is represented by the horizontal axis and total session cost is represented by the vertical axis. Plot the three given points on the graph. These three points appear to lie on a line. What information about the access plan suggests that the correct model is indeed a linear relationship?

Number of Minutes	Total Session Cost (in dollars)
0	
10	0.40
20	0.70
30	1.00
40	
50	
60	

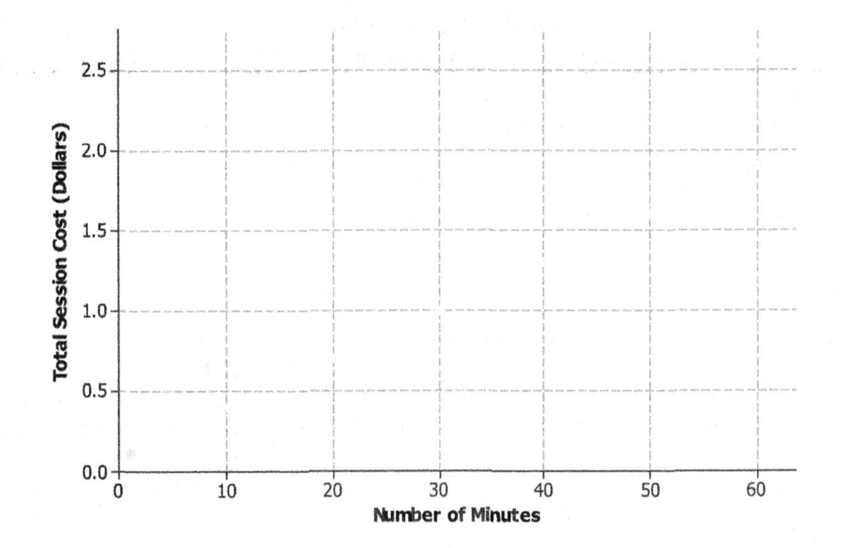

2. The rate of change describes how the total cost changes with respect to time.

   a. When the number of minutes increases by 10 (e.g., from 10 minutes to 20 minutes or from 20 minutes to 30 minutes), how much does the charge increase?

   b. Another way to say this would be the usage charge per 10 minutes of use. Use that information to determine the increase in cost based on only 1 minute of additional usage. In other words, find the usage charge per minute of use.

3. The company's pricing plan states that the usage rate is constant for any number of minutes connected to the Internet. In other words, the increase in cost for 10 more minutes of use (the value that you calculated in Exercise 2) is the same whether you increase from 20 to 30 minutes, 30 to 40 minutes, etc. Using this information, determine the total cost for 40 minutes, 50 minutes, and 60 minutes of use. Record those values in the table, and plot the corresponding points on the graph in Exercise 1.

4. Using the table and the graph in Exercise 1, compute the hypothetical cost for 0 minutes of use. What does that value represent in the context of the values that Lenore is trying to figure out?

5. On the graph in Exercise 1, draw a line through the points representing 0 to 60 minutes of use under this company's plan. The slope of this line is equal to the rate of change, which in this case is the usage rate.

6. Using x for the number of minutes and y for the total cost in dollars, write a function to model the linear relationship between minutes of use and total cost.

Example 2: Another Rate Plan

A second wireless access company has a similar method for computing its costs. Unlike the first company that Lenore was considering, this second company explicitly states its access fee is $\$0.15$, and its usage rate is $\$0.04$ per minute.

$$\text{Total Session Cost} = \$0.15 + \$0.04\ (\text{number of minutes})$$

Exercises 7–16

7. Let x represent the number of minutes used and y represent the total session cost in dollars. Construct a linear function that models the total session cost based on the number of minutes used.

8. Using the linear function constructed in Exercise 7, determine the total session cost for sessions of 0, 10, 20, 30, 40, 50, and 60 minutes, and fill in these values in the table below.

Number of Minutes	Total Session Cost (in dollars)
0	
10	
20	
30	
40	
50	
60	

9. Plot these points on the original graph in Exercise 1, and draw a line through these points. In what ways does the line that represents this second company's access plan differ from the line that represents the first company's access plan?

MP3 download sites are a popular forum for selling music. Different sites offer pricing that depends on whether or not you want to purchase an entire album or individual songs à la carte. One site offers MP3 downloads of individual songs with the following price structure: a $\$3$ fixed fee for a monthly subscription *plus* a charge of $\$0.25$ per song.

10. Using x for the number of songs downloaded and y for the total monthly cost in dollars, construct a linear function to model the relationship between the number of songs downloaded and the total monthly cost.

11. Using the linear function you wrote in Exercise 10, construct a table to record the total monthly cost (in dollars) for MP3 downloads of 10 songs, 20 songs, and so on up to 100 songs.

12. Plot the 10 data points in the table on a coordinate plane. Let the x-axis represent the number of songs downloaded and the y-axis represent the total monthly cost (in dollars) for MP3 downloads.

A band will be paid a flat fee for playing a concert. Additionally, the band will receive a fixed amount for every ticket sold. If 40 tickets are sold, the band will be paid $\$200$. If 70 tickets are sold, the band will be paid $\$260$.

13. Determine the rate of change.

14. Let x represent the number of tickets sold and y represent the amount the band will be paid in dollars. Construct a linear function to represent the relationship between the number of tickets sold and the amount the band will be paid.

15. What flat fee will the band be paid for playing the concert regardless of the number of tickets sold?

16. How much will the band receive for each ticket sold?

Lesson Summary

A linear function can be used to model a linear relationship between two types of quantities. The graph of a linear function is a straight line.

A linear function can be constructed using a rate of change and an initial value. It can be interpreted as an equation of a line in which:

- The rate of change is the slope of the line and describes how one quantity changes with respect to another quantity.
- The initial value is the y-intercept.

Problem Set

1. Recall that Lenore was investigating two wireless access plans. Her friend in Europe says that he uses a plan in which he pays a monthly fee of 30 euro plus 0.02 euro per minute of use.
   a. Construct a table of values for his plan's monthly cost based on 100 minutes of use for the month, 200 minutes of use, and so on up to $1{,}000$ minutes of use. (The charge of 0.02 euro per minute of use is equivalent to 2 euro per 100 minutes of use.)
   b. Plot these 10 points on a carefully labeled graph, and draw the line that contains these points.
   c. Let x represent minutes of use and y represent the total monthly cost in euro. Construct a linear function that determines monthly cost based on minutes of use.
   d. Use the function to calculate the cost under this plan for 750 minutes of use. If this point were added to the graph, would it be above the line, below the line, or on the line?

2. A shipping company charges a $\$4.45$ handling fee in addition to $\$0.27$ per pound to ship a package.
   a. Using x for the weight in pounds and y for the cost of shipping in dollars, write a linear function that determines the cost of shipping based on weight.
   b. Which line (solid, dotted, or dashed) on the following graph represents the shipping company's pricing method? Explain.

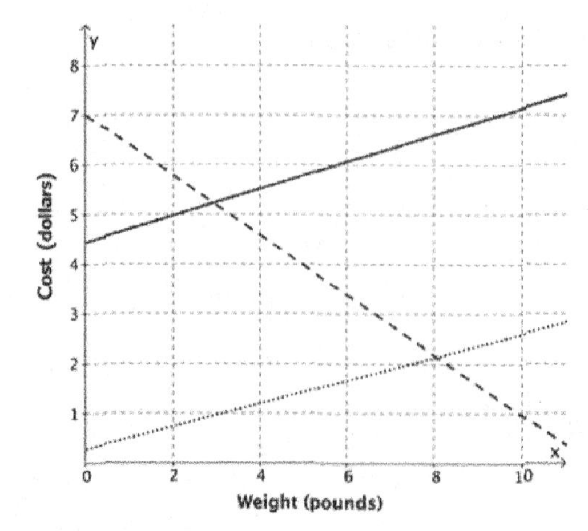

3. Kelly wants to add new music to her MP3 player. Another subscription site offers its downloading service using the following: Total Monthly Cost $= 5.25 + 0.30$ (number of songs).
   a. Write a sentence (all words, no math symbols) that the company could use on its website to explain how it determines the price for MP3 downloads for the month.
   b. Let x represent the number of songs downloaded and y represent the total monthly cost in dollars. Construct a function to model the relationship between the number of songs downloaded and the total monthly cost.
   c. Determine the cost of downloading 10 songs.

4. Li Na is saving money. Her parents gave her an amount to start, and since then she has been putting aside a fixed amount each week. After six weeks, Li Na has a total of $\$82$ of her own savings in addition to the amount her parents gave her. Fourteen weeks from the start of the process, Li Na has $\$118$.
   a. Using x for the number of weeks and y for the amount in savings (in dollars), construct a linear function that describes the relationship between the number of weeks and the amount in savings.
   b. How much did Li Na's parents give her to start?
   c. How much does Li Na set aside each week?
   d. Draw the graph of the linear function below (start by plotting the points for $x = 0$ and $x = 20$).

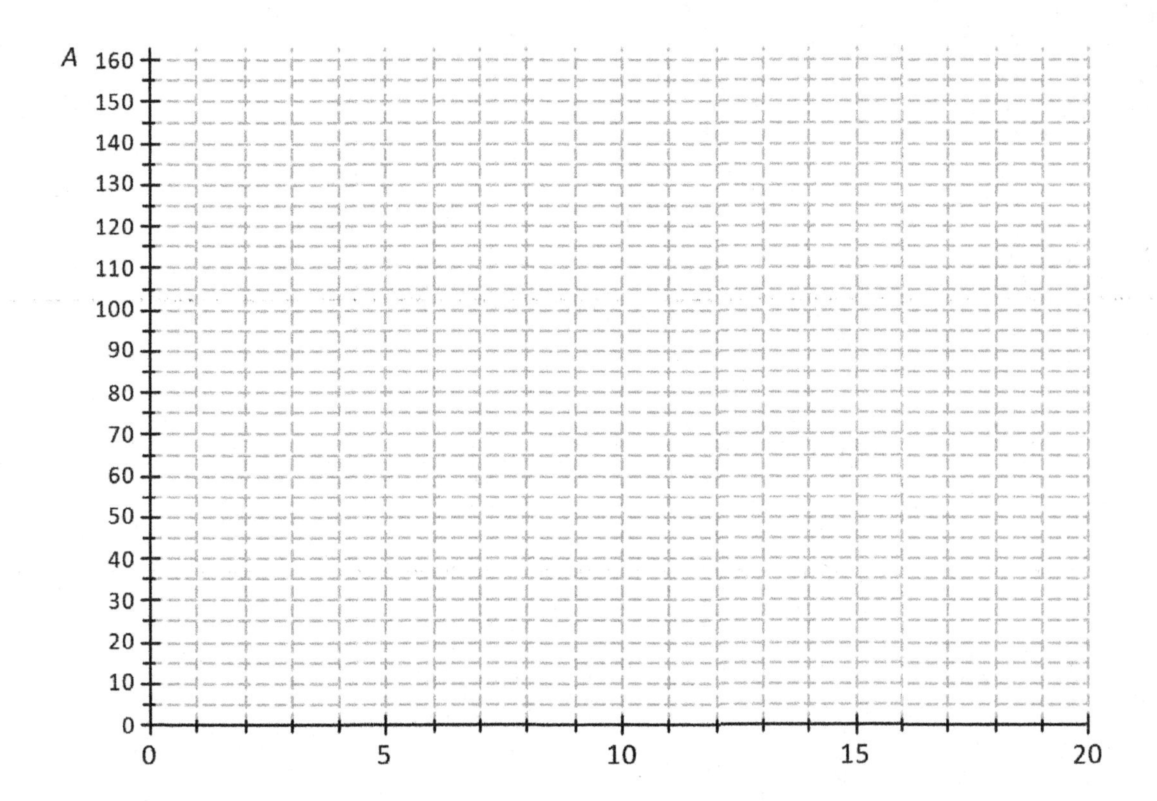

This page intentionally left blank

Lesson 2: Interpreting Rate of Change and Initial Value

Classwork

Linear functions are defined by the equation of a line. The graphs and the equations of the lines are important for understanding the relationship between the two variables represented in the following example as x and y.

Example 1: Rate of Change and Initial Value

The equation of a line can be interpreted as defining a linear function. The graphs and the equations of lines are important in understanding the relationship between two types of quantities (represented in the following examples by x and y).

In a previous lesson, you encountered an MP3 download site that offers downloads of individual songs with the following price structure: a $\$3$ fixed fee for a monthly subscription *plus* a fee of $\$0.25$ per song. The linear function that models the relationship between the number of songs downloaded and the total monthly cost of downloading songs can be written as

$$y = 0.25x + 3,$$

where x represents the number of songs downloaded and y represents the total monthly cost (in dollars) for MP3 downloads.

a. In your own words, explain the meaning of 0.25 within the context of the problem.

b. In your own words, explain the meaning of 3 within the context of the problem.

The values represented in the function can be interpreted in the following way:

$$y = \underbrace{0.25}_{\text{rate of change}}x + \underbrace{3}_{\text{initial value}}$$

The coefficient of x is referred to as the *rate of change*. It can be interpreted as the change in the values of y for every one-unit increase in the values of x. When the rate of change is positive, the linear function is *increasing*. In other words, *increasing* indicates that as the x-value increases, so does the y-value. When the rate of change is negative, the linear function is *decreasing*. *Decreasing* indicates that as the x-value increases, the y-value decreases.	The constant value is referred to as the *initial value* or y-intercept and can be interpreted as the value of y when $x = 0$.

Exercises 1–6: Is It a Better Deal?

Another site offers MP3 downloads with a different price structure: a $\$2$ fixed fee for a monthly subscription *plus* a fee of $\$0.40$ per song.

1. Write a linear function to model the relationship between the number of songs downloaded and the total monthly cost. As before, let x represent the number of songs downloaded and y represent the total monthly cost (in dollars) of downloading songs.

2. Determine the cost of downloading 0 songs and 10 songs from this site.

3. The graph below already shows the linear model for the first subscription site (Company 1): $y = 0.25x + 3$. Graph the equation of the line for the second subscription site (Company 2) by marking the two points from your work in Exercise 2 (for 0 songs and 10 songs) and drawing a line through those two points.

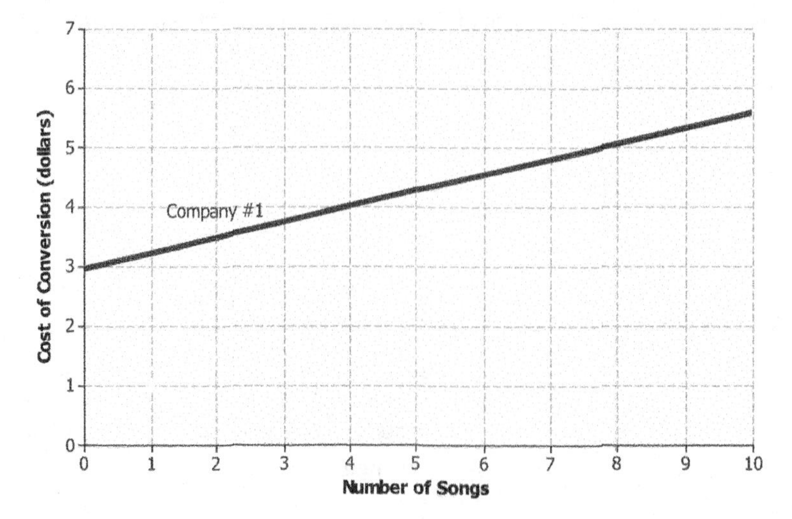

4. Which line has a steeper slope? Which company's model has the more expensive cost per song?

5. Which function has the greater initial value?

6. Which subscription site would you choose if you only wanted to download 5 songs per month? Which company would you choose if you wanted to download 10 songs? Explain your reasoning.

Exercises 7–9: Aging Autos

7. When someone purchases a new car and begins to drive it, the mileage (meaning the number of miles the car has traveled) immediately increases. Let x represent the number of years since the car was purchased and y represent the total miles traveled. The linear function that models the relationship between the number of years since purchase and the total miles traveled is $y = 15000x$.

   a. Identify and interpret the rate of change.

   b. Identify and interpret the initial value.

c. Is the mileage increasing or decreasing each year according to the model? Explain your reasoning.

8. When someone purchases a new car and begins to drive it, generally speaking, the resale value of the car (in dollars) goes down each year. Let x represent the number of years since purchase and y represent the resale value of the car (in dollars). The linear function that models the resale value based on the number of years since purchase is $y = 20000 - 1200x$.

   a. Identify and interpret the rate of change.

   b. Identify and interpret the initial value.

   c. Is the resale value increasing or decreasing each year according to the model? Explain.

9. Suppose you are given the linear function $y = 2.5x + 10$.

   a. Write a story that can be modeled by the given linear function.

   b. What is the rate of change? Explain its meaning with respect to your story.

   c. What is the initial value? Explain its meaning with respect to your story.

Lesson Summary

When a linear function is given by the equation of a line of the form $y = mx + b$, the rate of change is m, and the initial value is b. Both are easy to identify.

The rate of change of a linear function is the slope of the line it represents. It is the change in the values of y per a one-unit increase in the values of x.

- A positive rate of change indicates that a linear function is increasing.
- A negative rate of change indicates that a linear function is decreasing.
- Given two lines each with positive slope, the function represented by the steeper line has a greater rate of change.

The initial value of a linear function is the value of the y-variable when the x-value is zero.

Problem Set

1. A rental car company offers the following two pricing methods for its customers to choose from for a one-month rental:

   Method 1: Pay $\$400$ for the month, or

   Method 2: Pay $\$0.30$ per mile plus a standard maintenance fee of $\$35$.

   a. Construct a linear function that models the relationship between the miles driven and the total rental cost for Method 2. Let x represent the number of miles driven and y represent the rental cost (in dollars).

   b. If you plan to drive $1{,}100$ miles for the month, which method would you choose? Explain your reasoning.

2. Recall from a previous lesson that Kelly wants to add new music to her MP3 player. She was interested in a monthly subscription site that offered its MP3 downloading service for a monthly subscription fee *plus* a fee per song. The linear function that modeled the total monthly cost in dollars (y) based on the number of songs downloaded (x) is $y = 5.25 + 0.30x$.

   The site has suddenly changed its monthly price structure. The linear function that models the new total monthly cost in dollars (y) based on the number of songs downloaded (x) is $y = 0.35x + 4.50$.

   a. Explain the meaning of the value 4.50 in the new equation. Is this a better situation for Kelly than before?

   b. Explain the meaning of the value 0.35 in the new equation. Is this a better situation for Kelly than before?

   c. If you were to graph the two equations (old versus new), which line would have the steeper slope? What does this mean in the context of the problem?

   d. Which subscription plan provides the better value if Kelly downloads fewer than 15 songs per month?

This page intentionally left blank

Lesson 3: Representations of a Line

Classwork

Example 1: Rate of Change and Initial Value Given in the Context of the Problem

A truck rental company charges a $150 rental fee in addition to a charge of $0.50 per mile driven. Graph the linear function relating the total cost of the rental in dollars, C, to the number of miles driven, m, on the axes below.

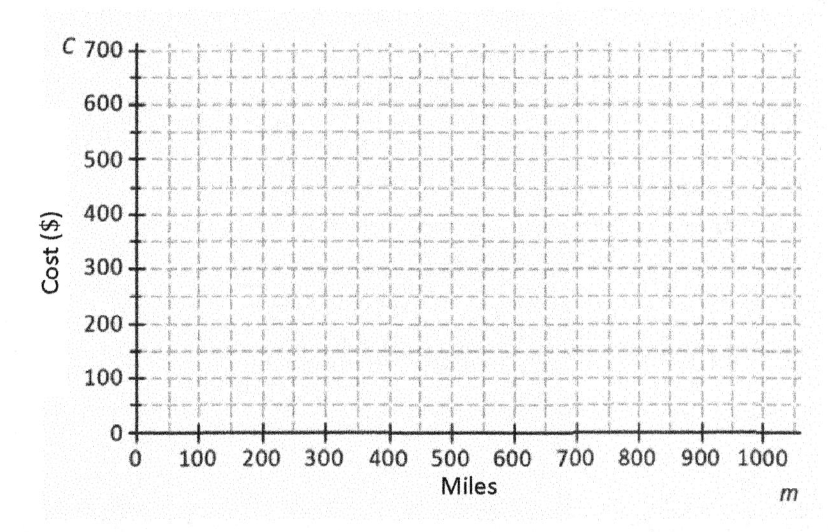

a. If the truck is driven 0 miles, what is the cost to the customer? How is this shown on the graph?

b. What is the rate of change that relates cost to number of miles driven? Explain what it means within the context of the problem.

c. On the axes given, sketch the graph of the linear function that relates C to m.

d. Write the equation of the linear function that models the relationship between number of miles driven and total rental cost.

Exercises

Jenna bought a used car for $\$18{,}000$. She has been told that the value of the car is likely to decrease by $\$2{,}500$ for each year that she owns the car. Let the value of the car in dollars be V and the number of years Jenna has owned the car be t.

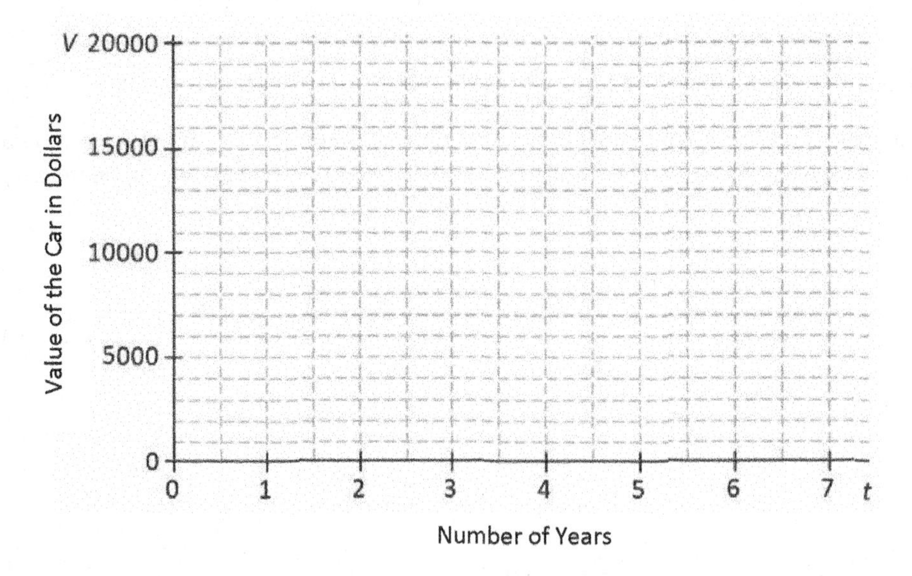

1. What is the value of the car when $t = 0$? Show this point on the graph.

2. What is the rate of change that relates V to t? (Hint: Is it positive or negative? How can you tell?)

3. Find the value of the car when:

   a. $t = 1$

   b. $t = 2$

   c. $t = 7$

4. Plot the points for the values you found in Exercise 3, and draw the line (using a straightedge) that passes through those points.

5. Write the linear function that models the relationship between the number of years Jenna has owned the car and the value of the car.

An online bookseller has a new book in print. The company estimates that if the book is priced at $15, then 800 copies of the book will be sold per day, and if the book is priced at $20, then 550 copies of the book will be sold per day.

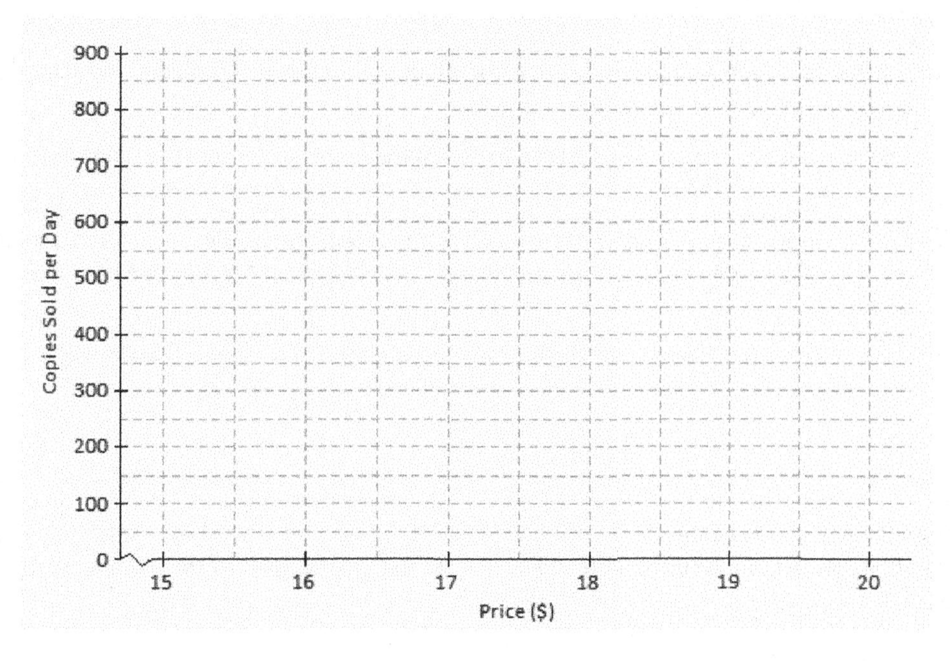

6. Identify the ordered pairs given in the problem. Then, plot both on the graph.

7. Assume that the relationship between the number of books sold and the price is linear. (In other words, assume that the graph is a straight line.) Using a straightedge, draw the line that passes through the two points.

8. What is the rate of change relating number of copies sold to price?

9. Based on the graph, if the company prices the book at $18, about how many copies of the book can they expect to sell per day?

10. Based on the graph, approximately what price should the company charge in order to sell 700 copies of the book per day?

Lesson Summary

When the rate of change, b, and an initial value, a, are given in the context of a problem, the linear function that models the situation is given by the equation $y = a + bx$.

The rate of change and initial value can also be used to sketch the graph of the linear function that models the situation.

When two or more ordered pairs are given in the context of a problem that involves a linear relationship, the graph of the linear function is the line that passes through those points. The linear function can be represented by the equation of that line.

Problem Set

1. A plumbing company charges a service fee of $\$120$, plus $\$40$ for each hour worked. Sketch the graph of the linear function relating the cost to the customer (in dollars), C, to the time worked by the plumber (in hours), t, on the axes below.

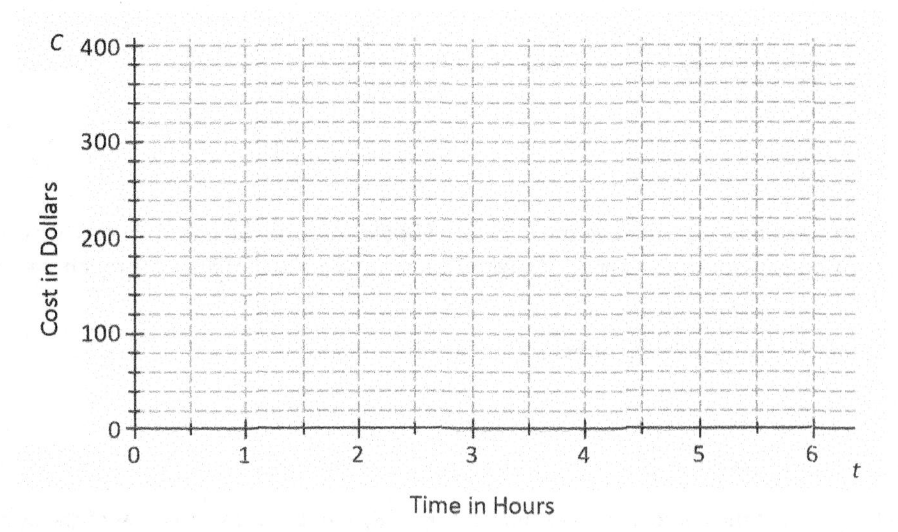

   a. If the plumber works for 0 hours, what is the cost to the customer? How is this shown on the graph?
   b. What is the rate of change that relates cost to time?
   c. Write a linear function that models the relationship between the hours worked and the cost to the customer.
   d. Find the cost to the customer if the plumber works for each of the following number of hours.
      i. 1 hour
      ii. 2 hours
      iii. 6 hours
   e. Plot the points for these times on the coordinate plane, and use a straightedge to draw the line through the points.

2. An author has been paid a writer's fee of $\$1,000$ plus $\$1.50$ for every copy of the book that is sold.

   a. Sketch the graph of the linear function that relates the total amount of money earned in dollars, A, to the number of books sold, n, on the axes below.

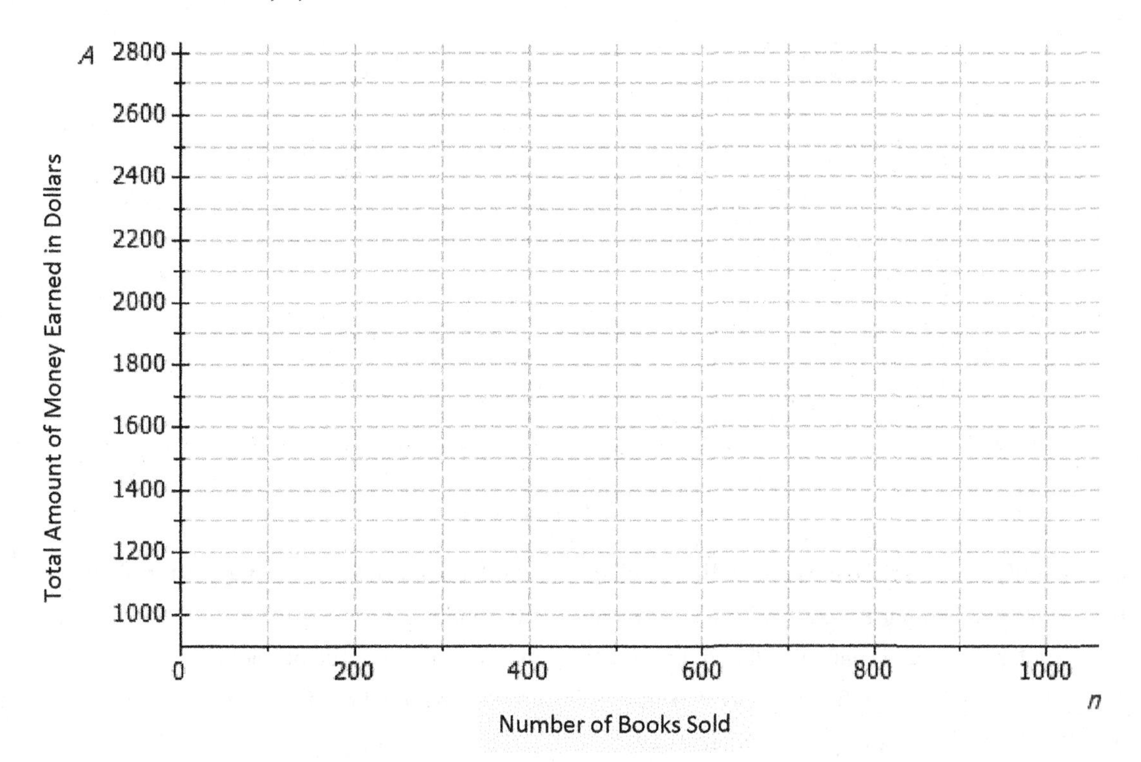

   b. What is the rate of change that relates the total amount of money earned to the number of books sold?

   c. What is the initial value of the linear function based on the graph?

   d. Let the number of books sold be n and the total amount earned be A. Construct a linear function that models the relationship between the number of books sold and the total amount earned.

3. Suppose that the price of gasoline has been falling. At the beginning of last month ($t = 0$), the price was $4.60 per gallon. Twenty days later ($t = 20$), the price was $4.20 per gallon. Assume that the price per gallon, P, fell at a constant rate over the twenty days.

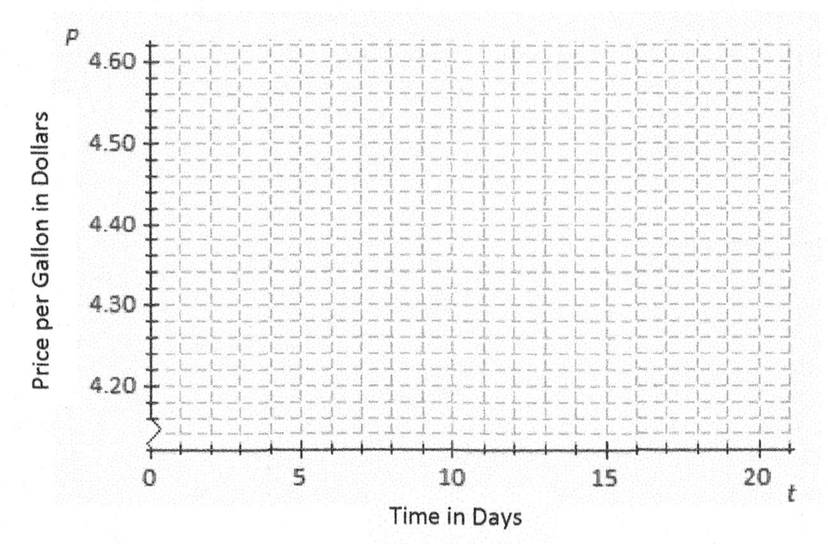

a. Identify the ordered pairs given in the problem. Plot both points on the coordinate plane above.

b. Using a straightedge, draw the line that contains the two points.

c. What is the rate of change? What does it mean within the context of the problem?

d. What is the function that models the relationship between the number of days and the price per gallon?

e. What was the price of gasoline after 9 days?

f. After how many days was the price $4.32?

Lesson 4: Increasing and Decreasing Functions

Classwork

Graphs are useful tools in terms of representing data. They provide a visual story, highlighting important facts that surround the relationship between quantities.

The graph of a linear function is a line. The slope of the line can provide useful information about the functional relationship between the two types of quantities:

- A linear function whose graph has a positive slope is said to be an *increasing function.*
- A linear function whose graph has a negative slope is said to be a *decreasing function.*
- A linear function whose graph has a zero slope is said to be a *constant function.*

Exercises

1. Read through each of the scenarios, and choose the graph of the function that best matches the situation. Explain the reason behind each choice.
   a. A bathtub is filled at a constant rate of 1.75 gallons per minute.
   b. A bathtub is drained at a constant rate of 2.5 gallons per minute.
   c. A bathtub contains 2.5 gallons of water.
   d. A bathtub is filled at a constant rate of 2.5 gallons per minute.

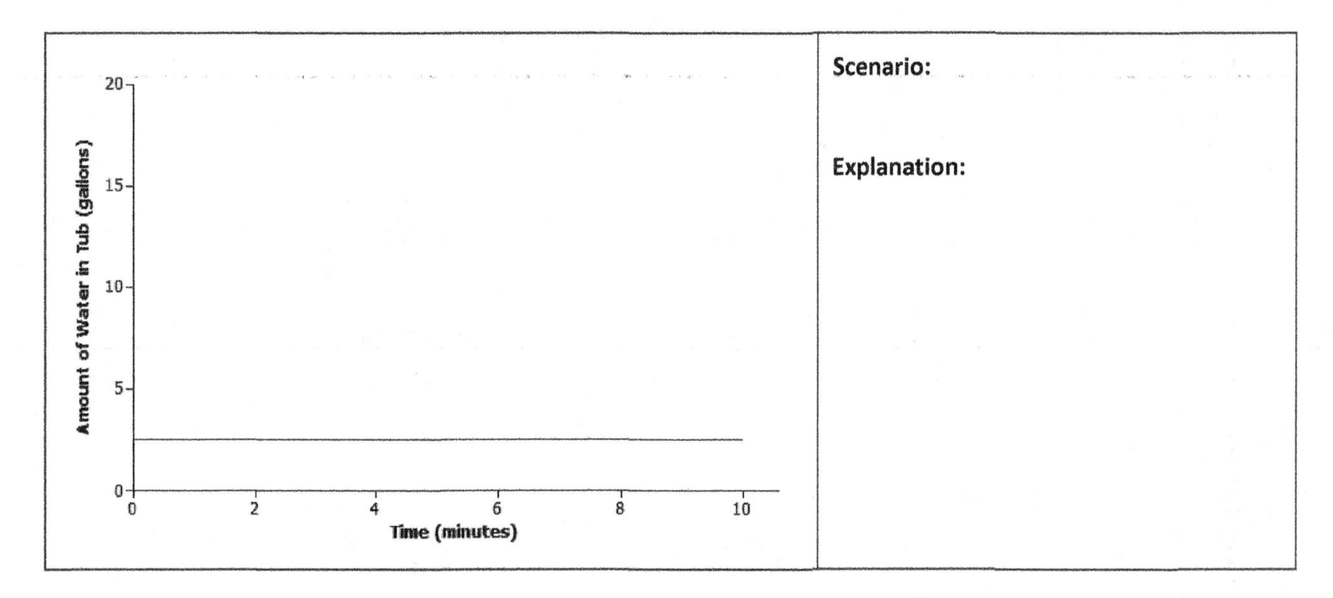

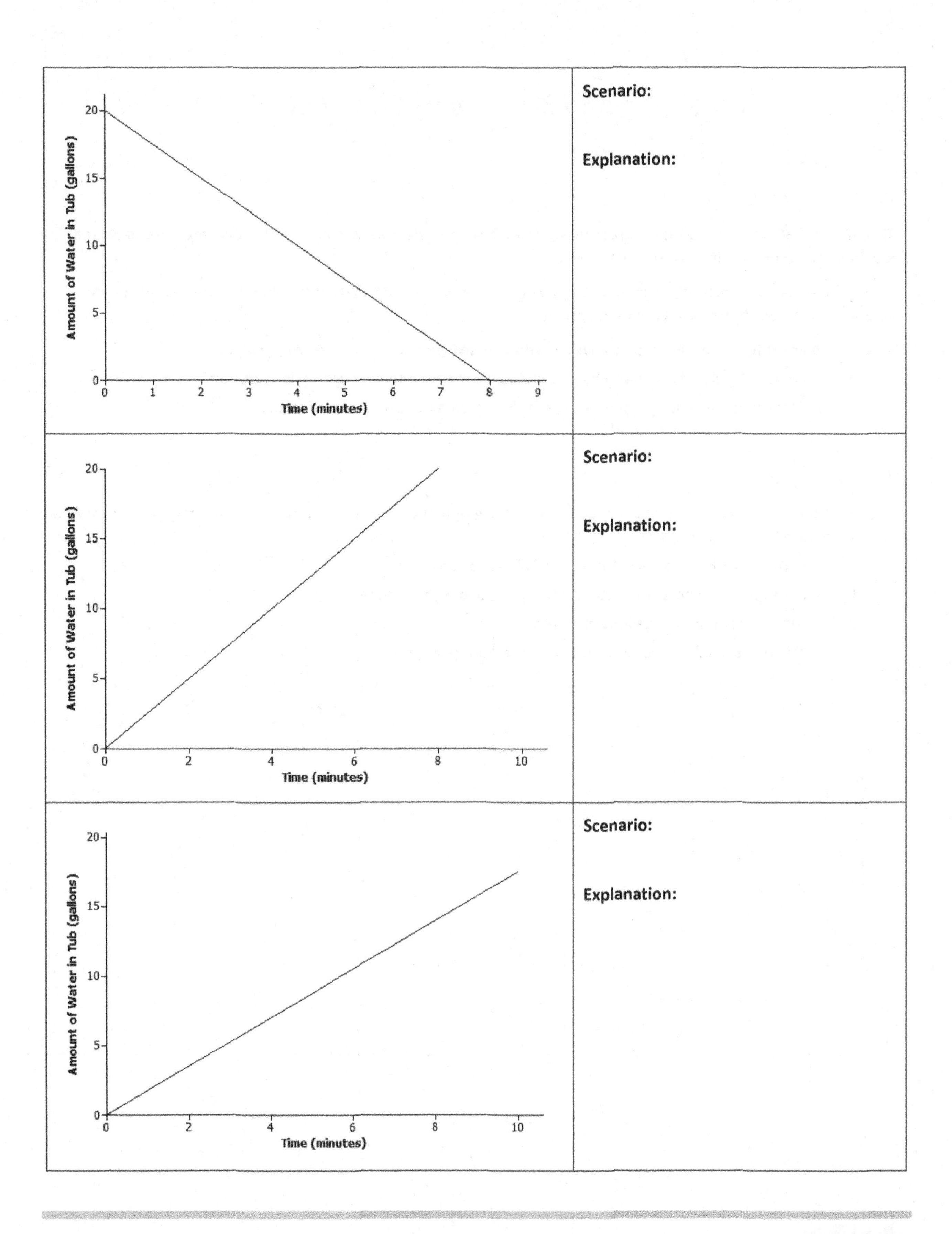

Amount of Water in Tub (gallons)
Time (minutes)
Scenario:
Explanation:
Amount of Water in Tub (gallons)
Time (minutes)
Scenario:
Explanation:
Amount of Water in Tub (gallons)
Time (minutes)
Scenario:
Explanation:

EUREKA MATH

2. Read through each of the scenarios, and sketch a graph of a function that models the situation.
   a. A messenger service charges a flat rate of $\$4.95$ to deliver a package regardless of the distance to the destination.

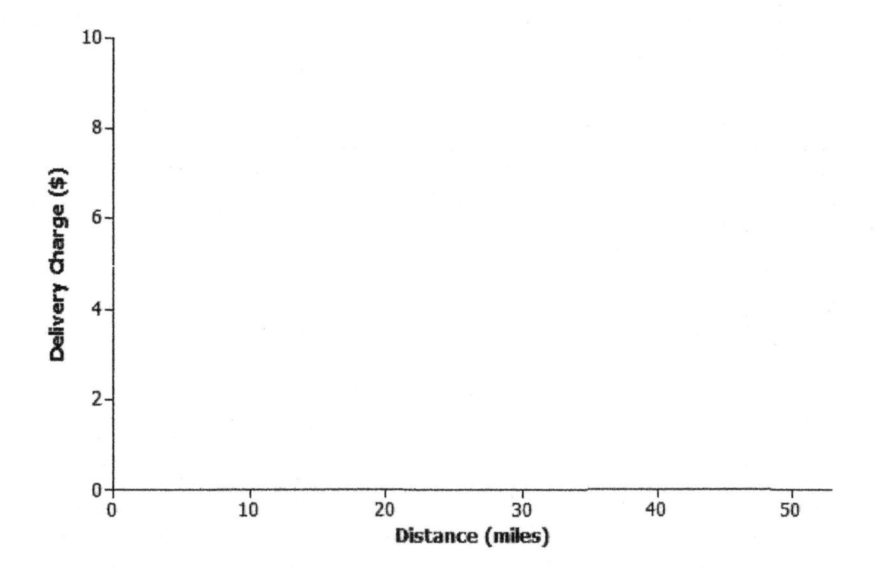

   b. At sea level, the air that surrounds us presses down on our bodies at 14.7 pounds per square inch (psi). For every 10 meters that you dive under water, the pressure increases by 14.7 psi.

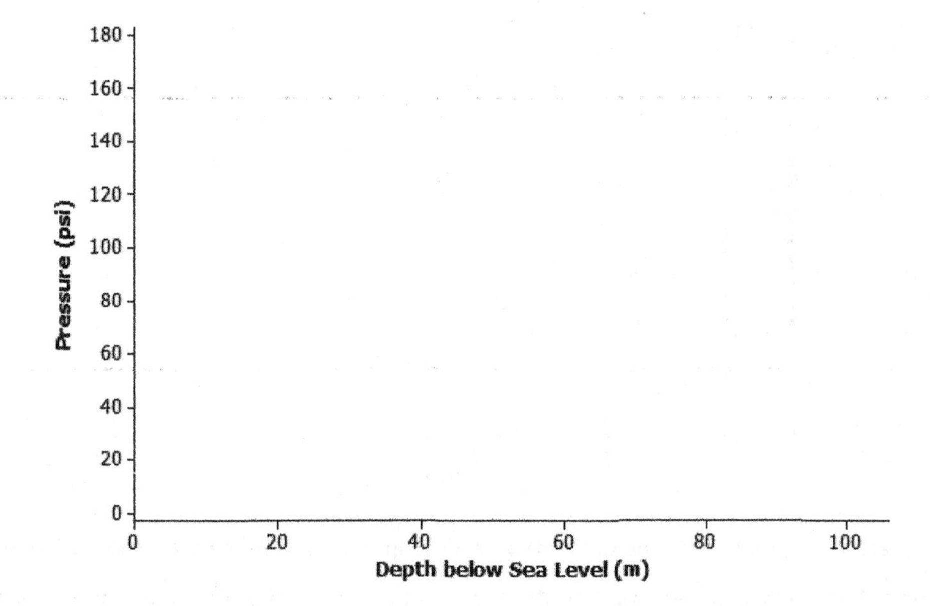

c. The range (driving distance per charge) of an electric car varies based on the average speed the car is driven. The initial range of the electric car after a full charge is 400 miles. However, the range is reduced by 20 miles for every 10 mph increase in average speed the car is driven.

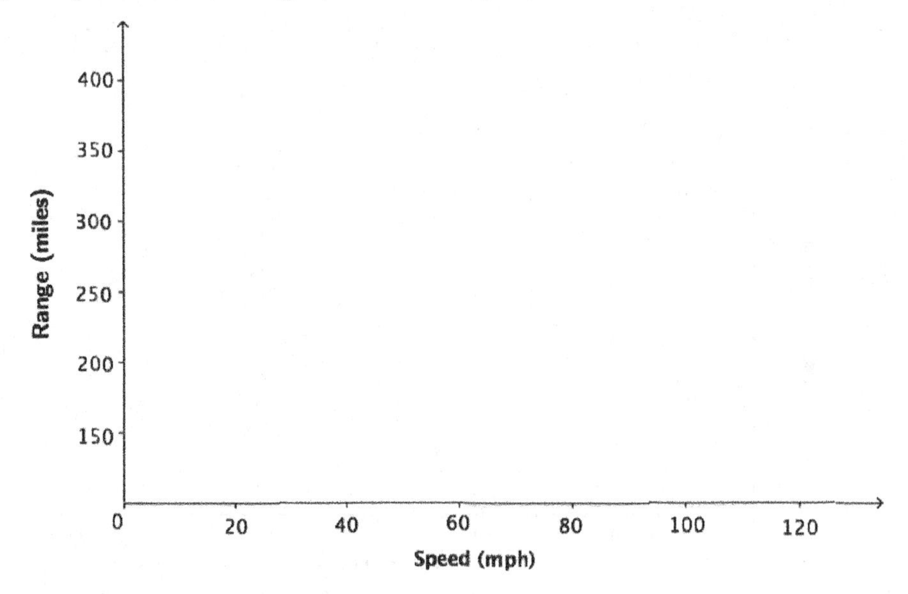

3. The graph below represents the total number of smartphones that are shipped to a retail store over the course of 50 days.

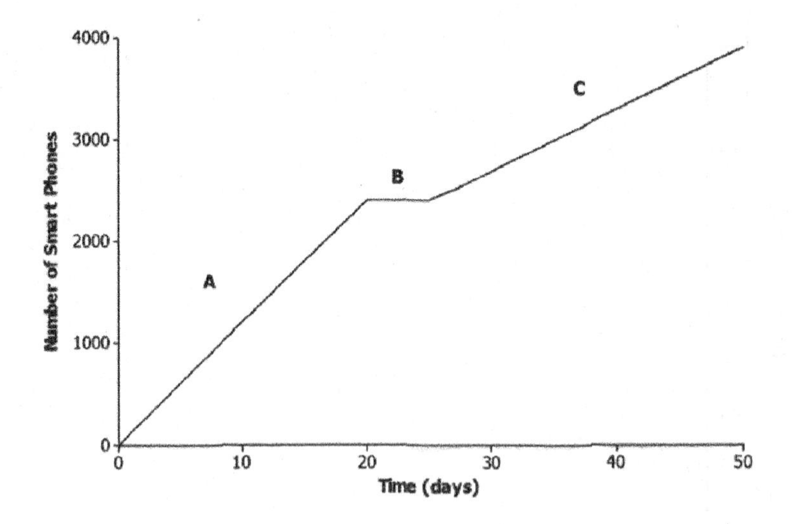

Match each part of the graph (A, B, and C) to its verbal description. Explain the reasoning behind your choice.

i. Half of the factory workers went on strike, and not enough smartphones were produced for normal shipments.

ii. The production schedule was normal, and smartphones were shipped to the retail store at a constant rate.

iii. A defective electronic chip was found, and the factory had to shut down, so no smartphones were shipped.

4. The relationship between Jameson's account balance and time is modeled by the graph below.

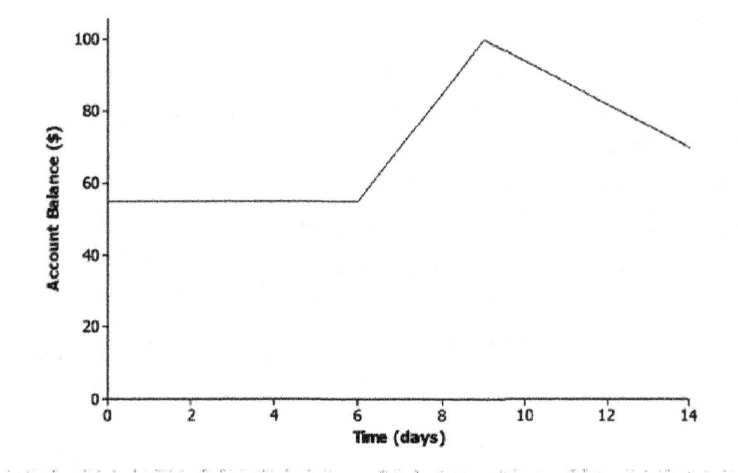

a. Write a story that models the situation represented by the graph.

b. When is the function represented by the graph increasing? How does this relate to your story?

c. When is the function represented by the graph decreasing? How does this relate to your story?

Lesson Summary

The graph of a function can be used to help describe the relationship between two types of quantities.

The slope of the line can provide useful information about the functional relationship between the quantities represented by the line:

- A function whose graph has a positive slope is said to be an *increasing function.*
- A function whose graph has a negative slope is said to be a *decreasing function.*
- A function whose graph has a zero slope is said to be a *constant function.*

Problem Set

1. Read through each of the scenarios, and choose the graph of the function that best matches the situation. Explain the reason behind each choice.
   a. The tire pressure on Regina's car remains at 30 psi.
   b. Carlita inflates her tire at a constant rate for 4 minutes.
   c. Air is leaking from Courtney's tire at a constant rate.

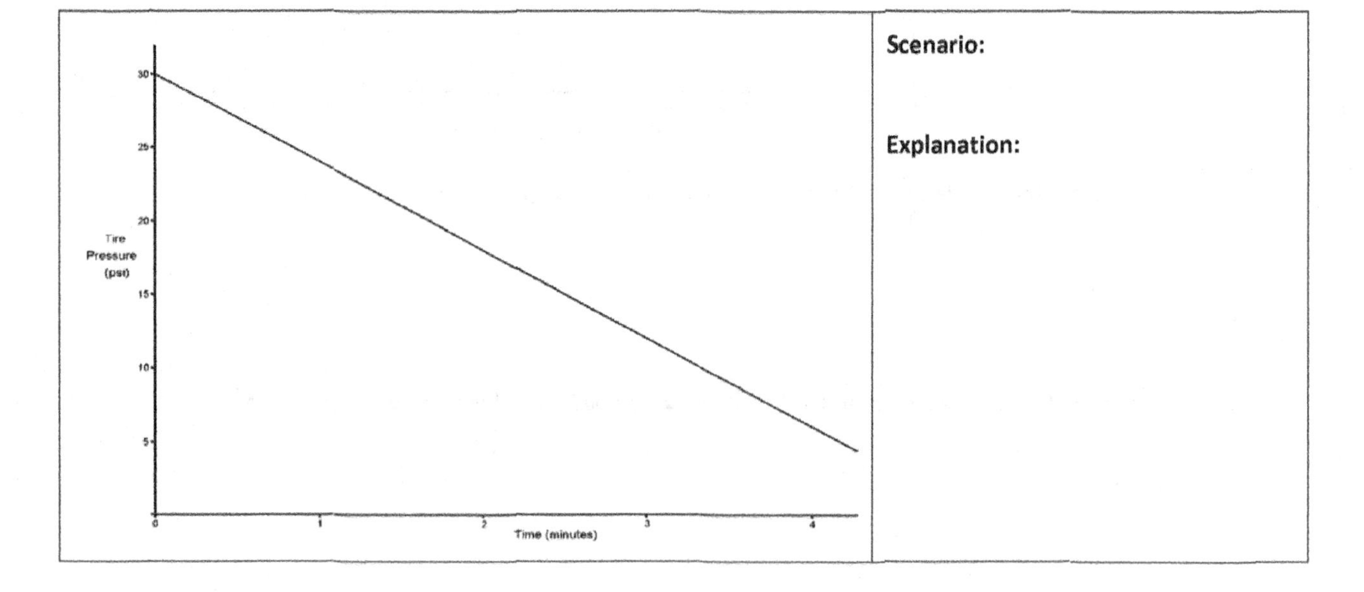

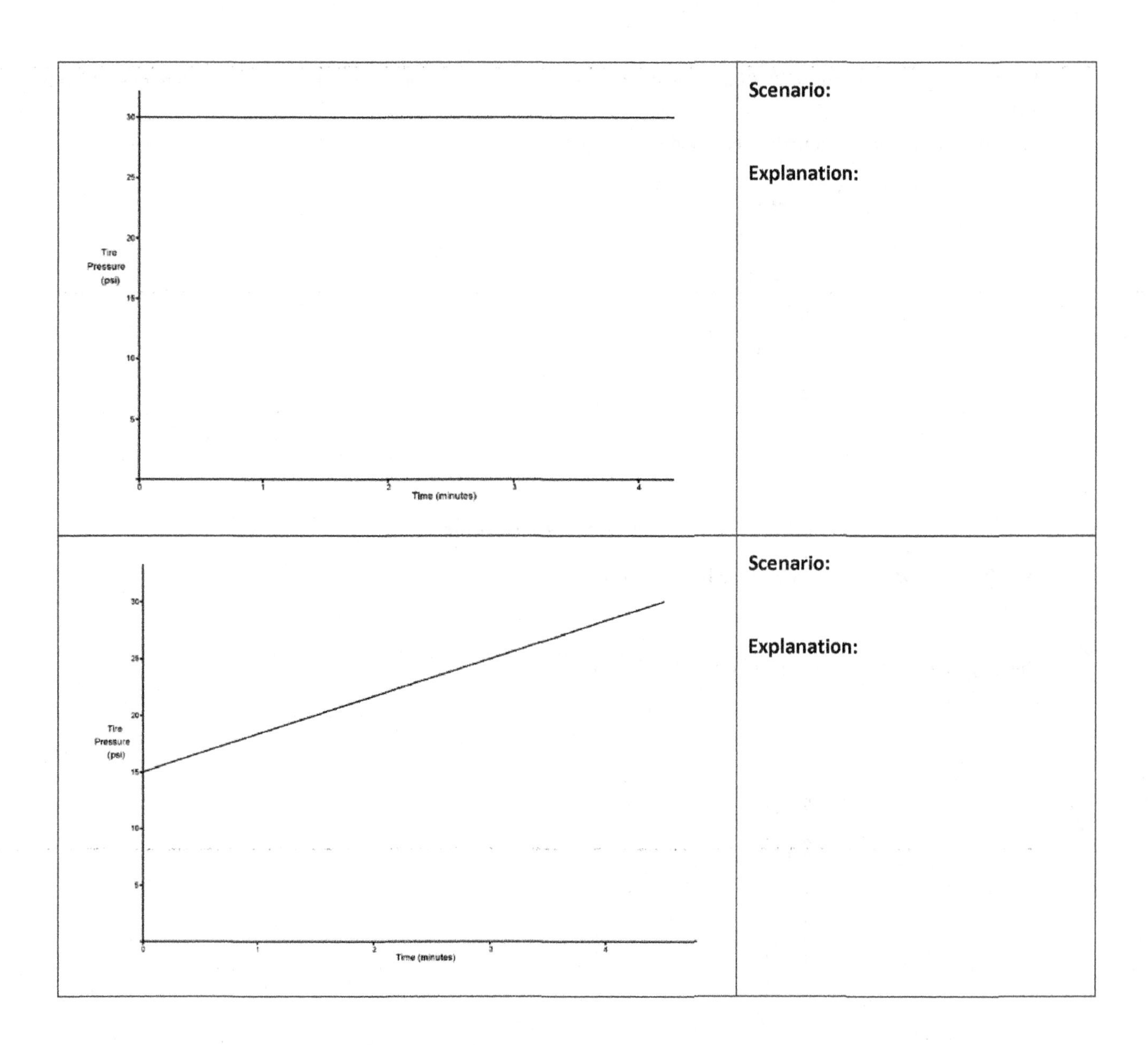

30
25
20
Tire Pressure (psi)
15
10
5
0
1
2
Time (minutes)
3
4
Scenario:
Explanation:
30
25
20
Tire Pressure (psi)
15
10
5
0
1
2
Time (minutes)
3
4
Scenario:
Explanation:

2. A home was purchased for $\$275{,}000$. Due to a recession, the value of the home fell at a constant rate over the next 5 years.

   a. Sketch a graph of a function that models the situation.

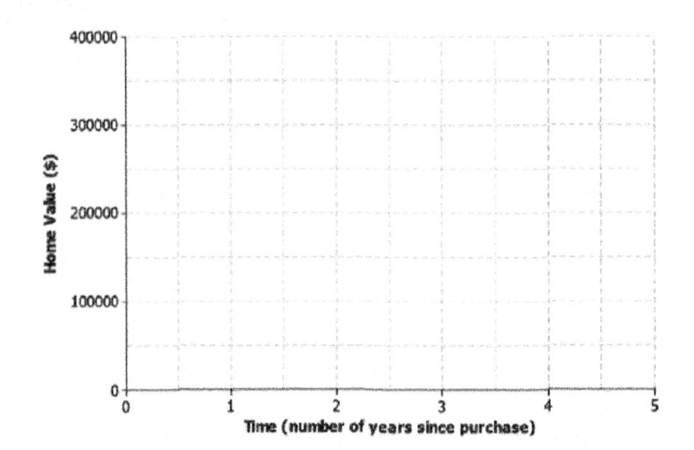

   b. Based on your graph, how is the home value changing with respect to time?

3. The graph below displays the first hour of Sam's bike ride.

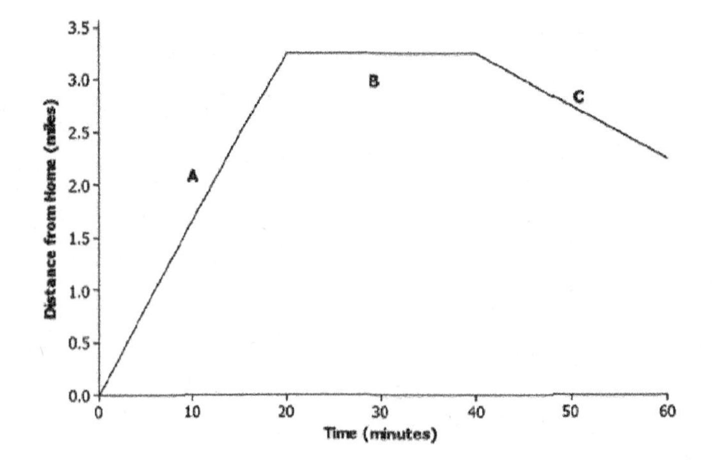

Match each part of the graph (A, B, and C) to its verbal description. Explain the reasoning behind your choice.

   i. Sam rides his bike to his friend's house at a constant rate.

   ii. Sam and his friend bike together to an ice cream shop that is between their houses.

   iii. Sam plays at his friend's house.

4. Using the axes below, create a story about the relationship between two quantities.

   a. Write a story about the relationship between two quantities. Any quantities can be used (e.g., distance and time, money and hours, age and growth). Be creative. Include keywords in your story such as *increase* and *decrease* to describe the relationship.

   b. Label each axis with the quantities of your choice, and sketch a graph of the function that models the relationship described in the story.

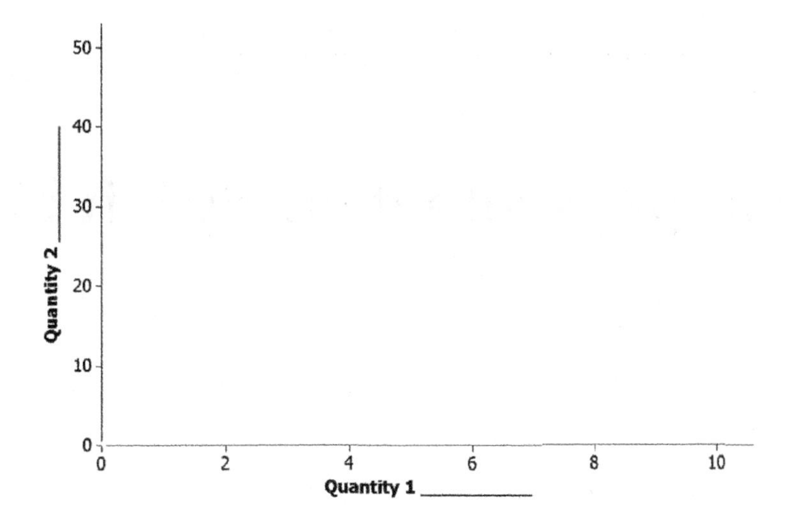

This page intentionally left blank

Lesson 5: Increasing and Decreasing Functions

Classwork

Example 1: Nonlinear Functions in the Real World

Not all real-world situations can be modeled by a linear function. There are times when a nonlinear function is needed to describe the relationship between two types of quantities. Compare the two scenarios:

a. Aleph is running at a constant rate on a flat, paved road. The graph below represents the total distance he covers with respect to time.

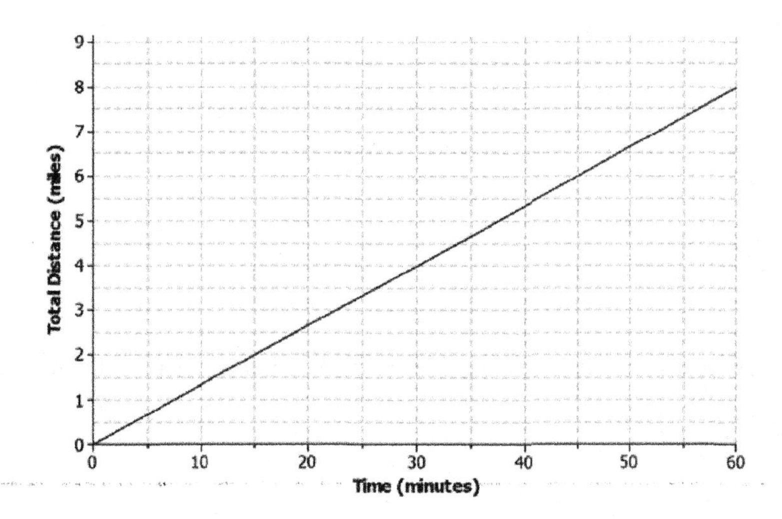

b. Shannon is running on a flat, rocky trail that eventually rises up a steep mountain. The graph below represents the total distance she covers with respect to time.

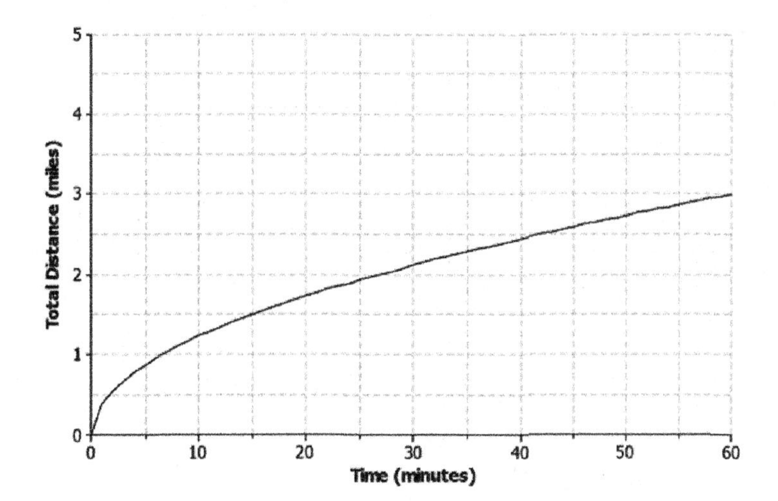

Exercises 1–2

1. In your own words, describe what is happening as Aleph is running during the following intervals of time.

   a. 0 to 15 minutes

   b. 15 to 30 minutes

   c. 30 to 45 minutes

   d. 45 to 60 minutes

2. In your own words, describe what is happening as Shannon is running during the following intervals of time.

   a. 0 to 15 minutes

   b. 15 to 30 minutes

   c. 30 to 45 minutes

   d. 45 to 60 minutes

Example 2: Increasing and Decreasing Functions

The rate of change of a function can provide useful information about the relationship between two quantities. A linear function has a constant rate of change. A nonlinear function has a variable rate of change.

Linear Functions	Nonlinear Functions
Linear function *increasing* at a constant rate 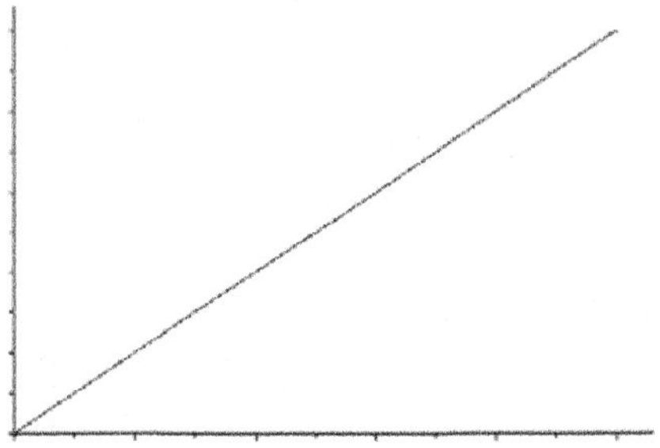	**Nonlinear function *increasing* at a variable rate** 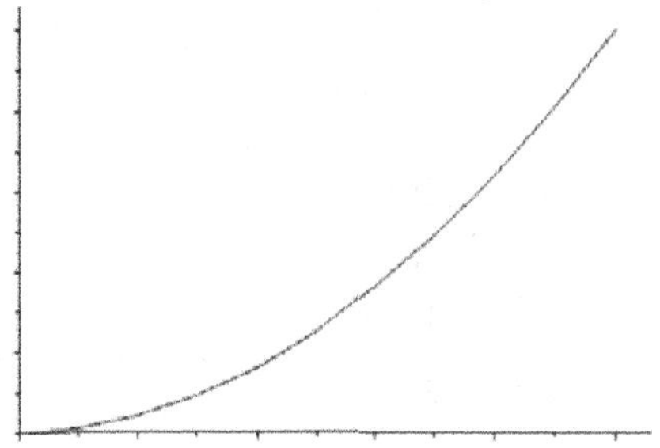
Linear function *decreasing* at a constant rate 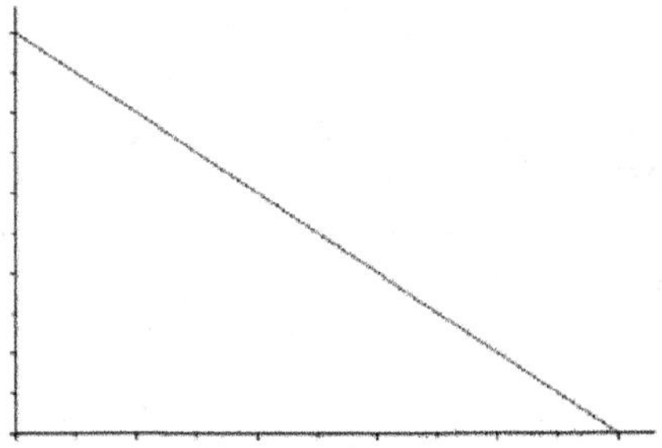	**Nonlinear function *decreasing* at a variable rate** 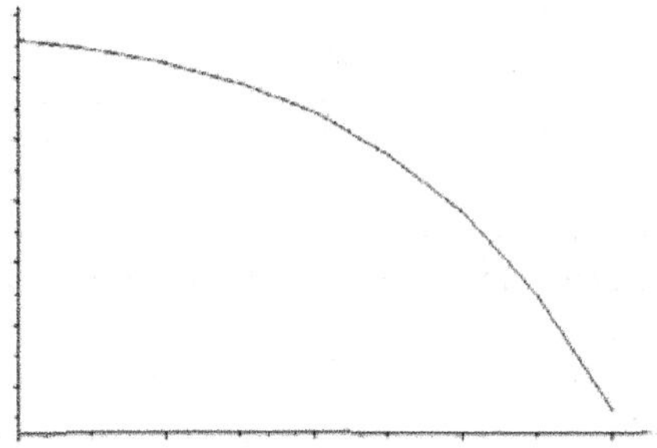
Linear function with a constant rate	**Nonlinear function with a variable rate**

Linear function with a constant rate:

x	y
0	7
1	10
2	13
3	16
4	19

Nonlinear function with a variable rate:

x	y
0	0
1	2
2	4
3	8
4	16

Exercises 3–5

3. Different breeds of dogs have different growth rates. A large breed dog typically experiences a rapid growth rate from birth to age 6 months. At that point, the growth rate begins to slow down until the dog reaches full growth around 2 years of age.

   a. Sketch a graph that represents the weight of a large breed dog from birth to 2 years of age.

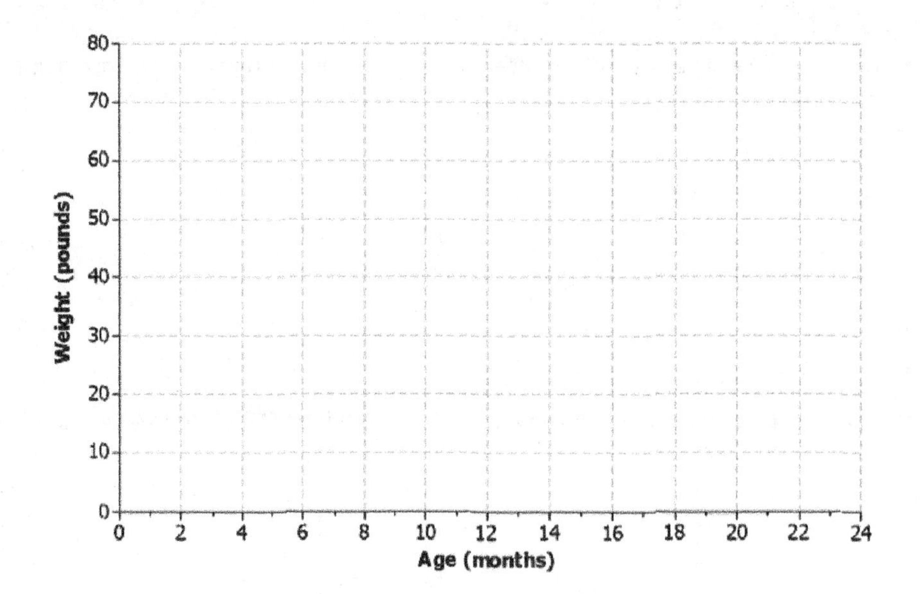

   b. Is the function represented by the graph linear or nonlinear? Explain.

   c. Is the function represented by the graph increasing or decreasing? Explain.

4. Nikka took her laptop to school and drained the battery while typing a research paper. When she returned home, Nikka connected her laptop to a power source, and the battery recharged at a constant rate.

   a. Sketch a graph that represents the battery charge with respect to time.

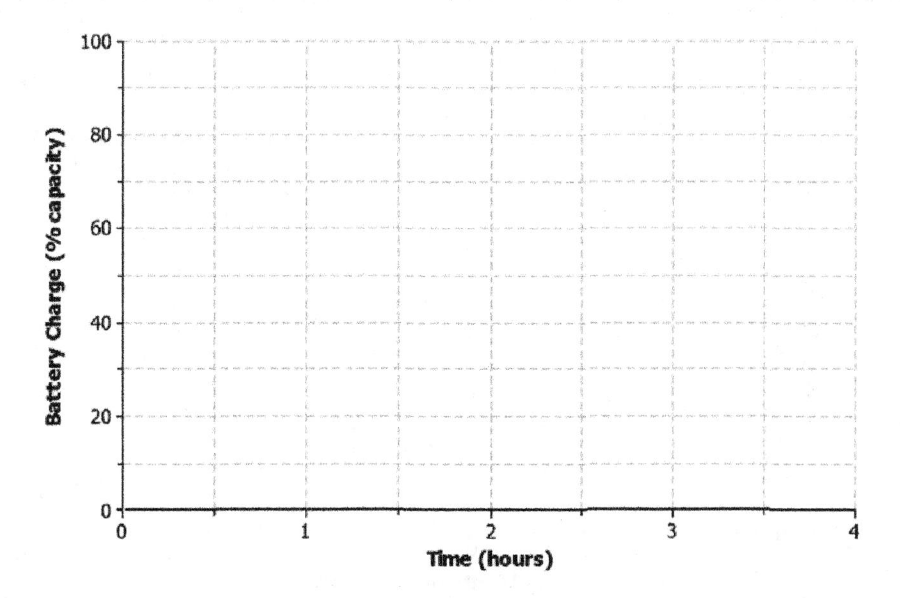

   b. Is the function represented by the graph linear or nonlinear? Explain.

   c. Is the function represented by the graph increasing or decreasing? Explain.

5. The long jump is a track-and-field event where an athlete attempts to leap as far as possible from a given point. Mike Powell of the United States set the long jump world record of 8.95 meters (29.4 feet) during the 1991 World Championships in Tokyo, Japan.

   a. Sketch a graph that represents the path of a high school athlete attempting the long jump.

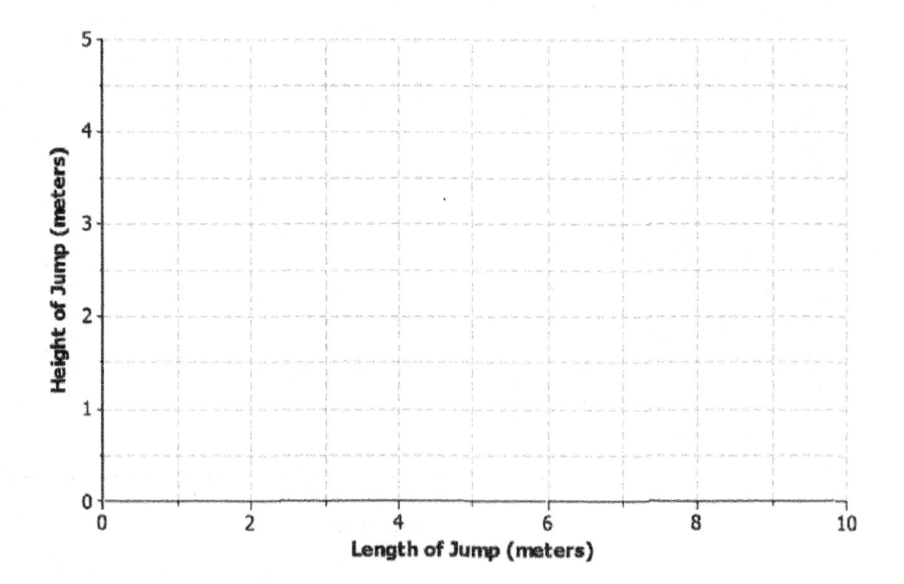

   b. Is the function represented by the graph linear or nonlinear? Explain.

   c. Is the function represented by the graph increasing or decreasing? Explain.

Example 3: Ferris Wheel

Lamar and his sister are riding a Ferris wheel at a state fair. Using their watches, they find that it takes 8 seconds for the Ferris wheel to make a complete revolution. The graph below represents Lamar and his sister's distance above the ground with respect to time.

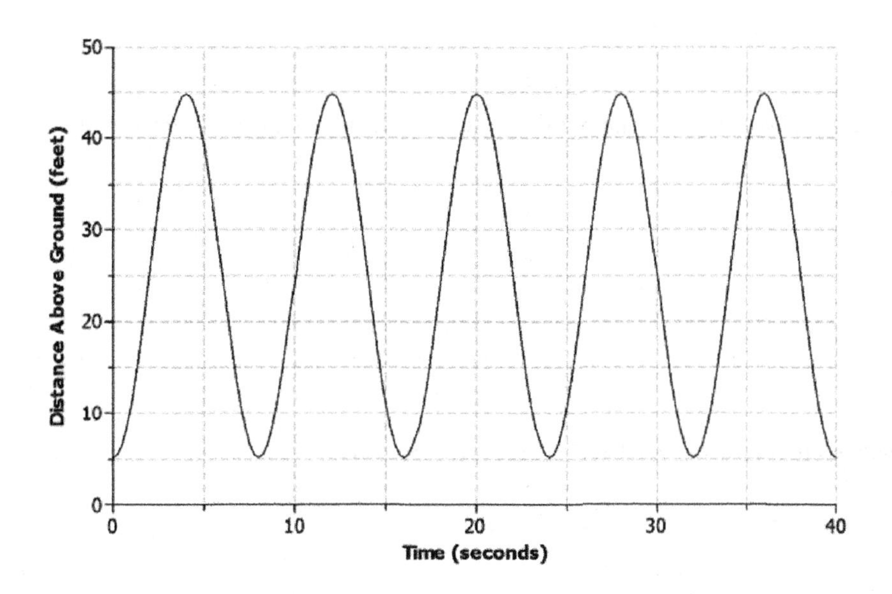

Exercises 6–9

6. Use the graph from Example 3 to answer the following questions.
   a. Is the function represented by the graph linear or nonlinear?

   b. Where is the function increasing? What does this mean within the context of the problem?

   c. Where is the function decreasing? What does this mean within the context of the problem?

7. How high above the ground is the platform for passengers to get on the Ferris wheel? Explain your reasoning.

8. Based on the graph, how many revolutions does the Ferris wheel complete during the 40-second time interval? Explain your reasoning.

9. What is the diameter of the Ferris wheel? Explain your reasoning.

Lesson Summary

The graph of a function can be used to help describe the relationship between the quantities it represents.

A linear function has a constant rate of change. A nonlinear function does not have a constant rate of change.

- A function whose graph has a positive rate of change is an *increasing function*.
- A function whose graph has a negative rate of change is a *decreasing function*.
- Some functions may increase and decrease over different intervals.

Problem Set

1. Read through the following scenarios, and match each to its graph. Explain the reasoning behind your choice.
   a. This shows the change in a smartphone battery charge as a person uses the phone more frequently.
   b. A child takes a ride on a swing.
   c. A savings account earns simple interest at a constant rate.
   d. A baseball has been hit at a Little League game.

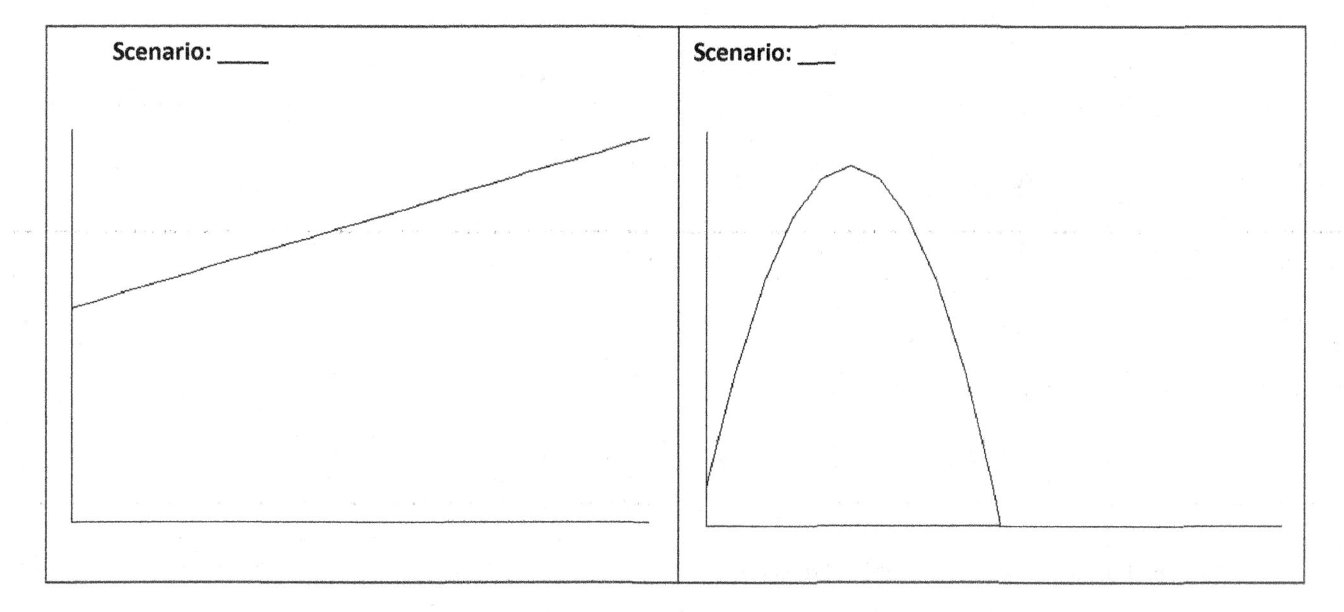

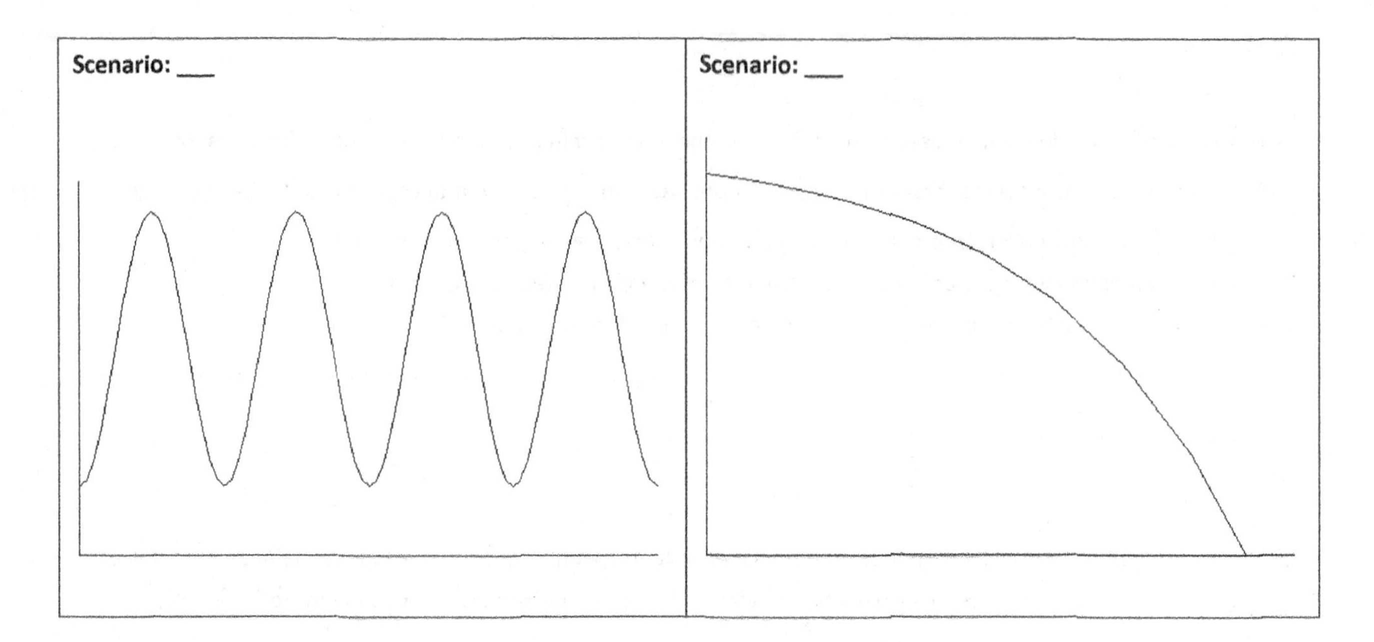

2. The graph below shows the volume of water for a given creek bed during a 24-hour period. On this particular day, there was wet weather with a period of heavy rain.

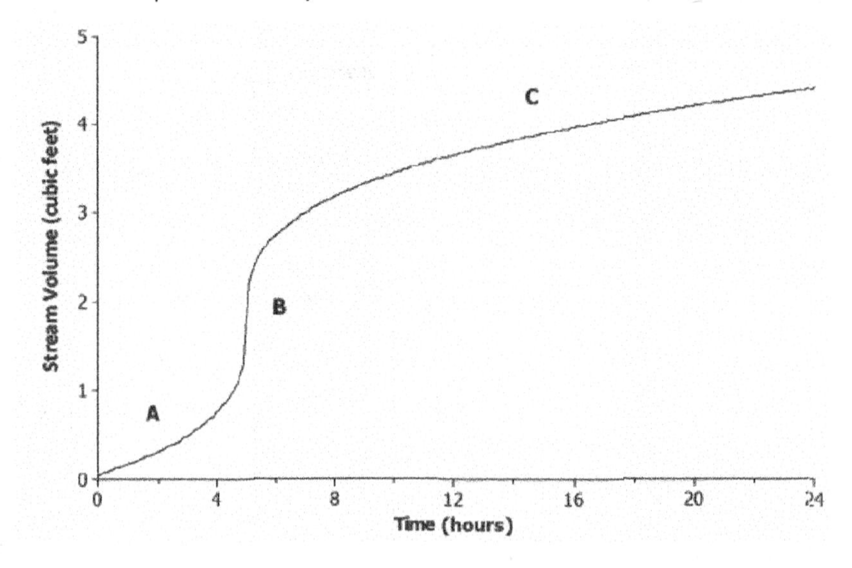

Describe how each part (A, B, and C) of the graph relates to the scenario.

3. Half-life is the time required for a quantity to fall to half of its value measured at the beginning of the time period. If there are 100 grams of a radioactive element to begin with, there will be 50 grams after the first half-life, 25 grams after the second half-life, and so on.

   a. Sketch a graph that represents the amount of the radioactive element left with respect to the number of half-lives that have passed.

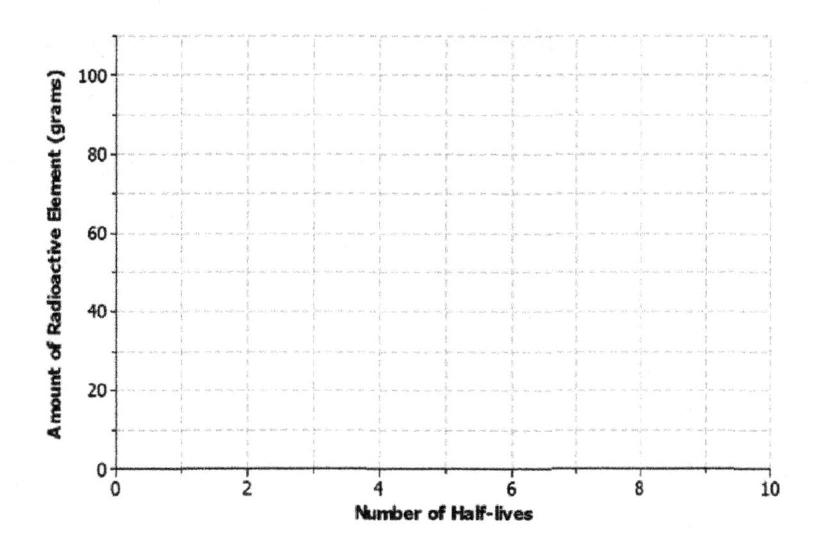

   b. Is the function represented by the graph linear or nonlinear? Explain.

   c. Is the function represented by the graph increasing or decreasing?

4. Lanae parked her car in a no-parking zone. Consequently, her car was towed to an impound lot. In order to release her car, she needs to pay the impound lot charges. There is an initial charge on the day the car is brought to the lot. However, 10% of the previous day's charges will be added to the total charge for every day the car remains in the lot.

   a. Sketch a graph that represents the total charges with respect to the number of days a car remains in the impound lot.

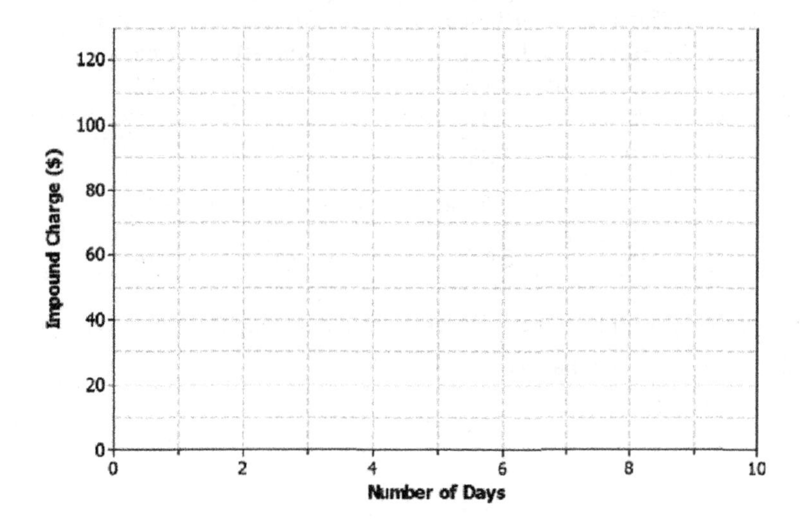

b. Is the function represented by the graph linear or nonlinear? Explain.

c. Is the function represented by the graph increasing or decreasing? Explain.

5. Kern won a $50 gift card to his favorite coffee shop. Every time he visits the shop, he purchases the same coffee drink.

a. Sketch a graph of a function that can be used to represent the amount of money that remains on the gift card with respect to the number of drinks purchased.

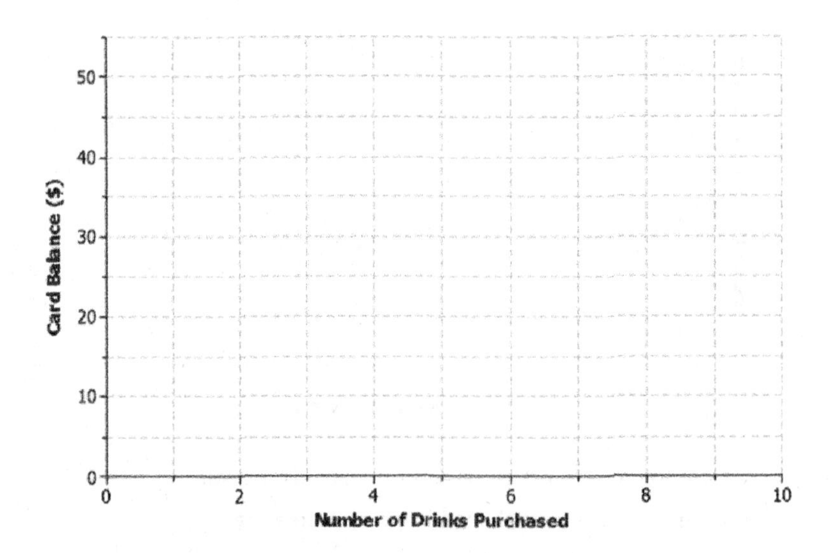

b. Is the function represented by the graph linear or nonlinear? Explain.

c. Is the function represented by the graph increasing or decreasing? Explain.

6. Jay and Brooke are racing on bikes to a park 8 miles away. The tables below display the total distance each person biked with respect to time.

Jay

Time (minutes)	Distance (miles)
0	0
5	0.84
10	1.86
15	3.00
20	4.27
25	5.67

Brooke

Time (minutes)	Distance (miles)
0	0
5	1.2
10	2.4
15	3.6
20	4.8
25	6.0

a. Which person's biking distance could be modeled by a nonlinear function? Explain.

b. Who would you expect to win the race? Explain.

7. Using the axes in Problem 7(b), create a story about the relationship between two quantities.
   a. Write a story about the relationship between two quantities. Any quantities can be used (e.g., distance and time, money and hours, age and growth). Be creative! Include keywords in your story such as *increase* and *decrease* to describe the relationship.
   b. Label each axis with the quantities of your choice, and sketch a graph of the function that models the relationship described in the story.

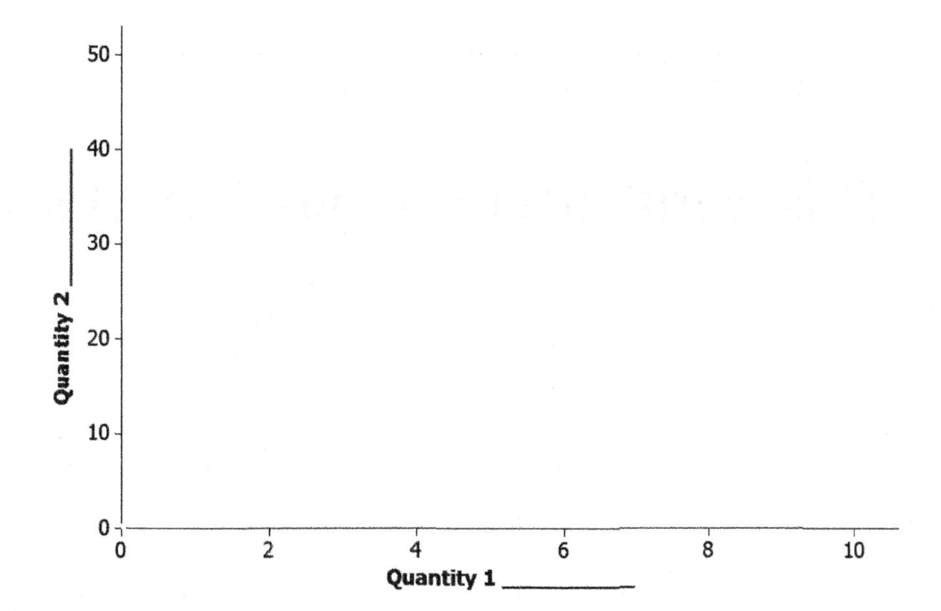

This page intentionally left blank

Lesson 6: Scatter Plots

Classwork

Example 1

A bivariate data set consists of observations on two variables. For example, you might collect data on 13 different car models. Each observation in the data set would consist of an (x, y) pair.

x: weight (in pounds, rounded to the nearest 50 pounds)

and

y: fuel efficiency (in miles per gallon, mpg)

The table below shows the weight and fuel efficiency for 13 car models with automatic transmissions manufactured in 2009 by Chevrolet.

Model	Weight (pounds)	Fuel Efficiency (mpg)
1	3,200	23
2	2,550	28
3	4,050	19
4	4,050	20
5	3,750	20
6	3,550	22
7	3,550	19
8	3,500	25
9	4,600	16
10	5,250	12
11	5,600	16
12	4,500	16
13	4,800	15

Exercises 1–8

1. In the Example 1 table, the observation corresponding to Model 1 is $(3200, 23)$. What is the fuel efficiency of this car? What is the weight of this car?

2. Add the points corresponding to the other 12 observations to the scatter plot.

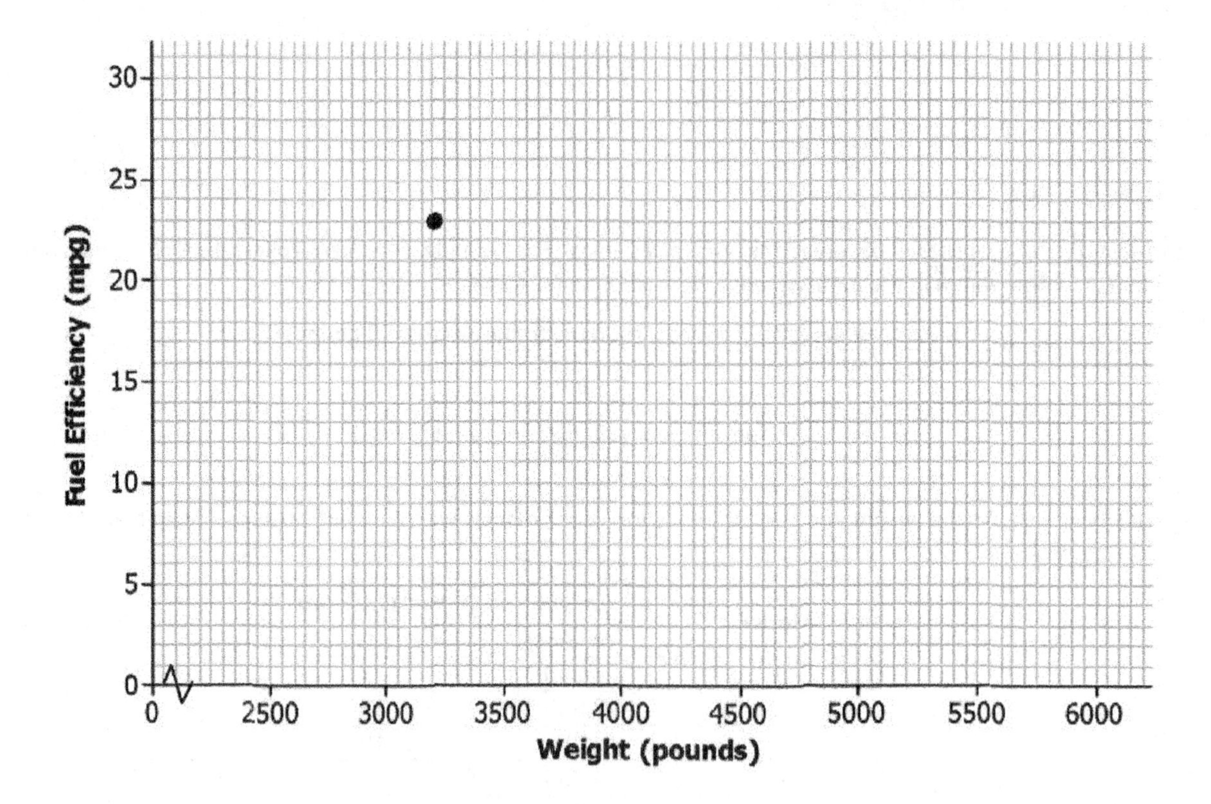

3. Do you notice a pattern in the scatter plot? What does this imply about the relationship between weight (x) and fuel efficiency (y)?

Is there a relationship between price and the quality of athletic shoes? The data in the table below are from the *Consumer Reports* website.

x: price (in dollars)

and

y: *Consumer Reports* quality rating

The quality rating is on a scale of 0 to 100, with 100 being the highest quality.

Shoe	Price (dollars)	Quality Rating
1	65	71
2	45	70
3	45	62
4	80	59
5	110	58
6	110	57
7	30	56
8	80	52
9	110	51
10	70	51

4. One observation in the data set is $(110, 57)$. What does this ordered pair represent in terms of cost and quality?

5. To construct a scatter plot of these data, you need to start by thinking about appropriate scales for the axes of the scatter plot. The prices in the data set range from \$30 to \$110, so one reasonable choice for the scale of the x-axis would range from \$20 to \$120, as shown below. What would be a reasonable choice for a scale for the y-axis?

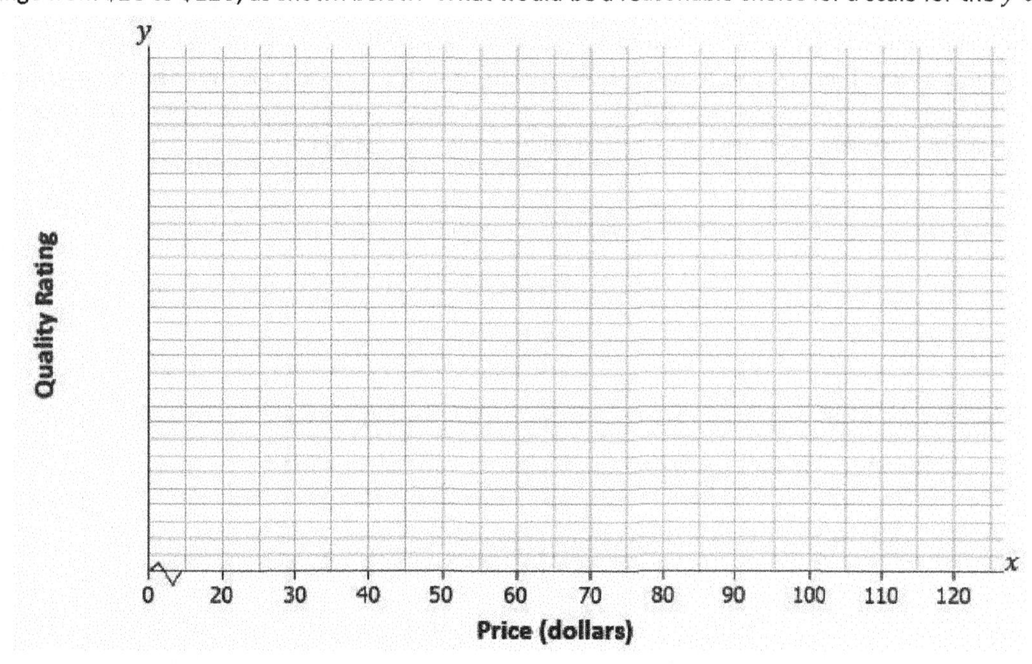

6. Add a scale to the y-axis. Then, use these axes to construct a scatter plot of the data.

7. Do you see any pattern in the scatter plot indicating that there is a relationship between price and quality rating for athletic shoes?

8. Some people think that if shoes have a high price, they must be of high quality. How would you respond?

Example 2: Statistical Relationships

A pattern in a scatter plot indicates that the values of one variable tend to vary in a predictable way as the values of the other variable change. This is called a *statistical relationship*. In the fuel efficiency and car weight example, fuel efficiency tended to decrease as car weight increased.

This is useful information, but be careful not to jump to the conclusion that increasing the weight of a car *causes* the fuel efficiency to go down. There may be some other explanation for this. For example, heavier cars may also have bigger engines, and bigger engines may be less efficient. You cannot conclude that changes to one variable *cause* changes in the other variable just because there is a statistical relationship in a scatter plot.

Exercises 9–10

9. Data were collected on

x: shoe size

and

y: score on a reading ability test

for 29 elementary school students. The scatter plot of these data is shown below. Does there appear to be a statistical relationship between shoe size and score on the reading test?

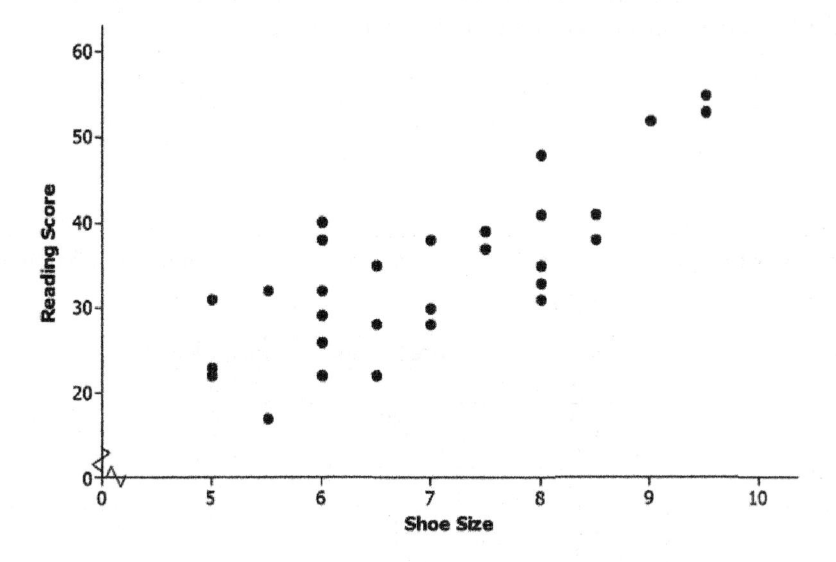

10. Explain why it is not reasonable to conclude that having big feet causes a high reading score. Can you think of a different explanation for why you might see a pattern like this?

Lesson Summary

- A scatter plot is a graph of numerical data on two variables.
- A pattern in a scatter plot suggests that there may be a relationship between the two variables used to construct the scatter plot.
- If two variables tend to vary together in a predictable way, we can say that there is a statistical relationship between the two variables.
- A statistical relationship between two variables does not imply that a change in one variable causes a change in the other variable (a cause-and-effect relationship).

Problem Set

1. The table below shows the price and overall quality rating for 15 different brands of bike helmets.

   Data source: www.consumerreports.org

Helmet	Price (dollars)	Quality Rating
A	35	65
B	20	61
C	30	60
D	40	55
E	50	54
F	23	47
G	30	47
H	18	43
I	40	42
J	28	41
K	20	40
L	25	32
M	30	63
N	30	63
O	40	53

Construct a scatter plot of price (x) and quality rating (y). Use the grid below.

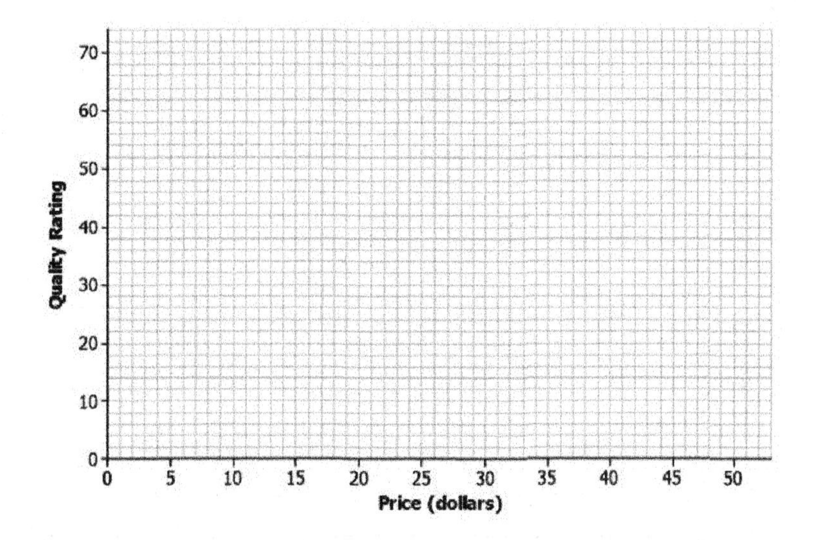

2. Do you think that there is a statistical relationship between price and quality rating? If so, describe the nature of the relationship.

3. Scientists are interested in finding out how different species adapt to finding food sources. One group studied crocodilian species to find out how their bite force was related to body mass and diet. The table below displays the information they collected on body mass (in pounds) and bite force (in pounds).

Species	Body Mass (pounds)	Bite Force (pounds)
Dwarf crocodile	35	450
Crocodile F	40	260
Alligator A	30	250
Caiman A	28	230
Caiman B	37	240
Caiman C	45	255
Crocodile A	110	550
Nile crocodile	275	650
Crocodile B	130	500
Crocodile C	135	600
Crocodile D	135	750
Caiman D	125	550
Indian Gharial crocodile	225	400
Crocodile G	220	1,000
American croc	270	900
Crocodile E	285	750
Crocodile F	425	1,650
American alligator	300	1,150
Alligator B	325	1,200
Alligator C	365	1,450

Data Source: http://journals.plos.org/plosone/article?id=10.1371/journal.pone.0031781#pone-0031781-t001

(Note: Body mass and bite force have been converted to pounds from kilograms and newtons, respectively.)

Construct a scatter plot of body mass (x) and bite force (y). Use the grid below, and be sure to add an appropriate scale to the axes.

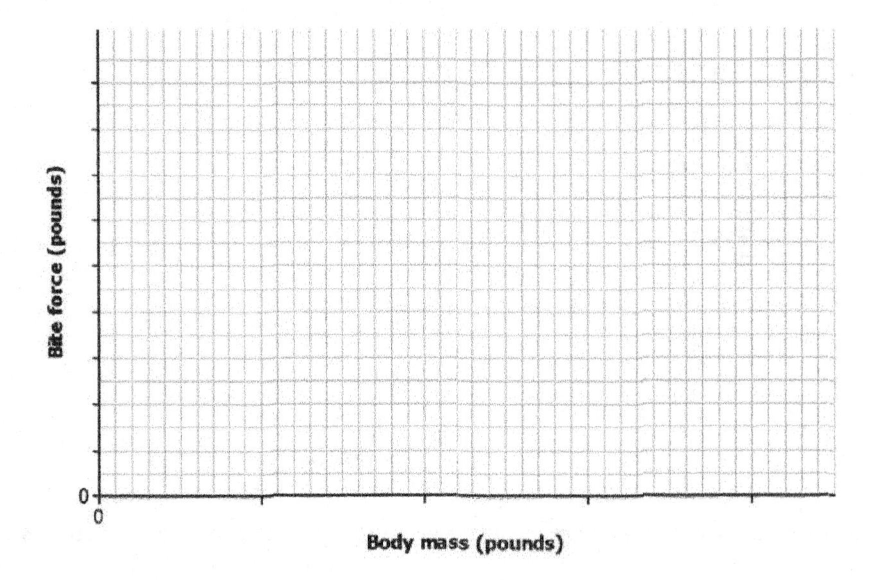

4. Do you think that there is a statistical relationship between body mass and bite force? If so, describe the nature of the relationship.

5. Based on the scatter plot, can you conclude that increased body mass causes increased bite force? Explain.

Lesson 7: Patterns in Scatter Plots

Classwork

Example 1

In the previous lesson, you learned that scatter plots show trends in bivariate data.

When you look at a scatter plot, you should ask yourself the following questions:

a. Does it look like there is a relationship between the two variables used to make the scatter plot?

b. If there is a relationship, does it appear to be linear?

c. If the relationship appears to be linear, is the relationship a positive linear relationship or a negative linear relationship?

To answer the first question, look for patterns in the scatter plot. Does there appear to be a general pattern to the points in the scatter plot, or do the points look as if they are scattered at random? If you see a pattern, you can answer the second question by thinking about whether the pattern would be well described by a line. Answering the third question requires you to distinguish between a positive linear relationship and a negative linear relationship. A positive linear relationship is one that is described by a line with a positive slope. A negative linear relationship is one that is described by a line with a negative slope.

Exercises 1–9

Take a look at the following five scatter plots. Answer the three questions in Example 1 for each scatter plot.

1. Scatter Plot 1

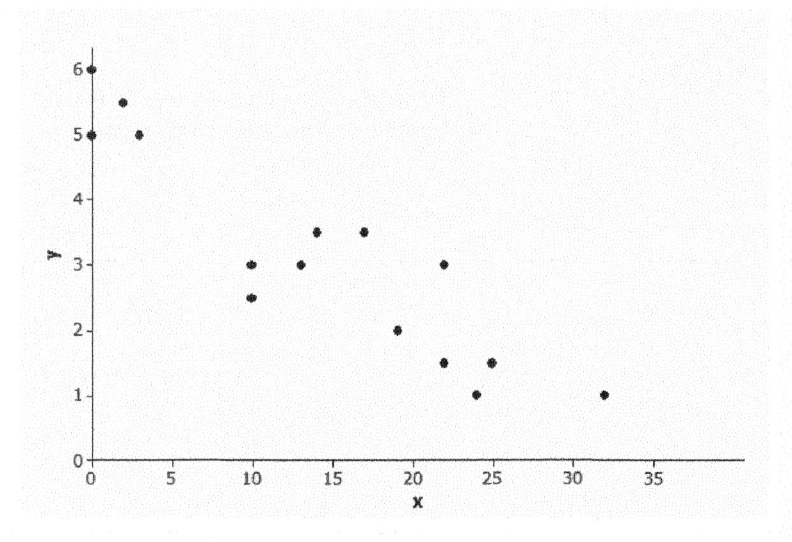

Is there a relationship?

If there is a relationship, does it appear to be linear?

If the relationship appears to be linear, is it a positive or a negative linear relationship?

2. Scatter Plot 2

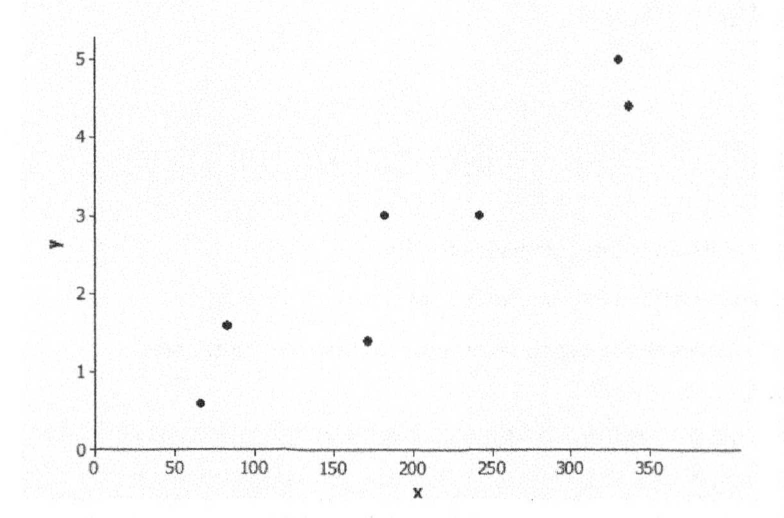

Is there a relationship?

If there is a relationship, does it appear to be linear?

If the relationship appears to be linear, is it a positive or a negative linear relationship?

3. Scatter Plot 3

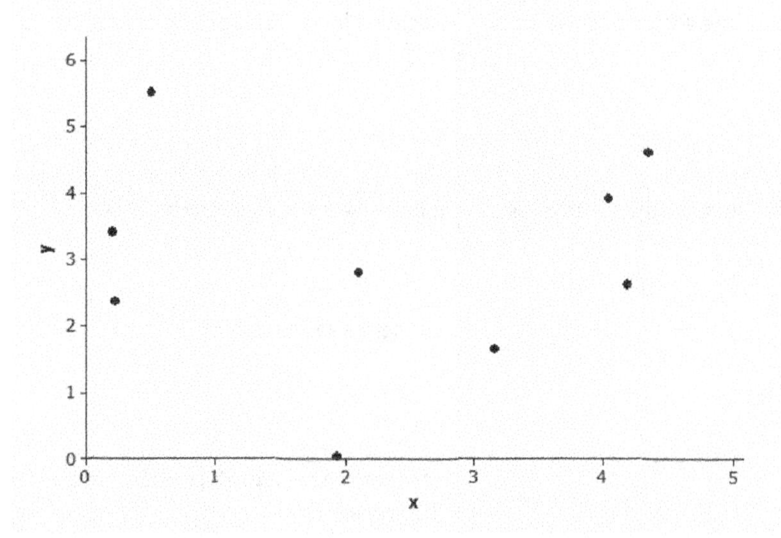

Is there a relationship?

If there is a relationship, does it appear to be linear?

If the relationship appears to be linear, is it a positive or a negative linear relationship?

4. Scatter Plot 4

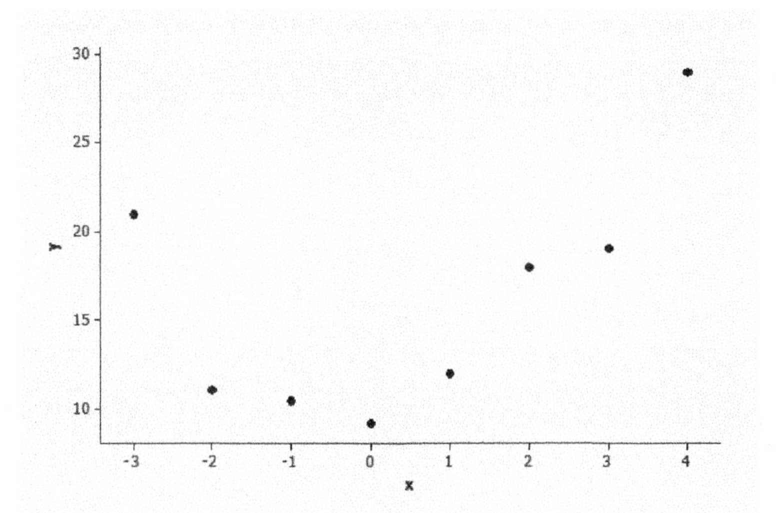

Is there a relationship?

If there is a relationship, does it appear to be linear?

If the relationship appears to be linear, is it a positive or a negative linear relationship?

5. Scatter Plot 5

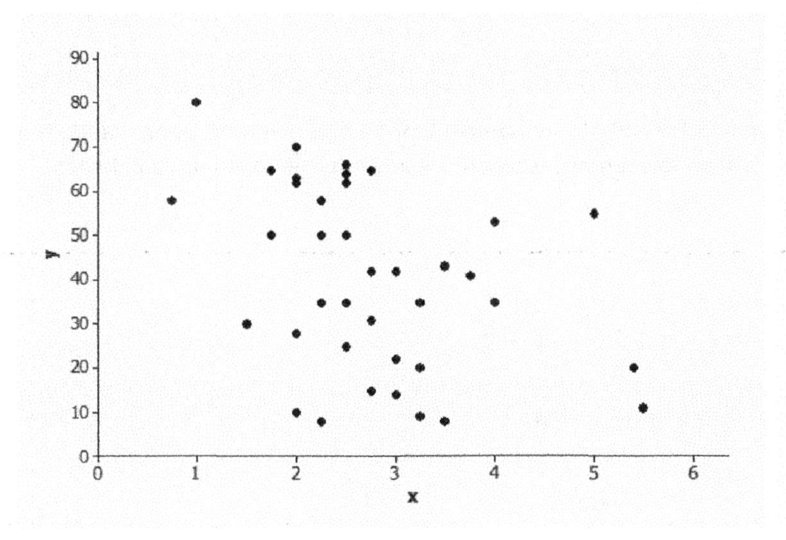

Is there a relationship?

If there is a relationship, does it appear to be linear?

If the relationship appears to be linear, is it a positive or a negative linear relationship?

6. Below is a scatter plot of data on weight in pounds (x) and fuel efficiency in miles per gallon (y) for 13 cars. Using the questions at the beginning of this lesson as a guide, write a few sentences describing any possible relationship between x and y.

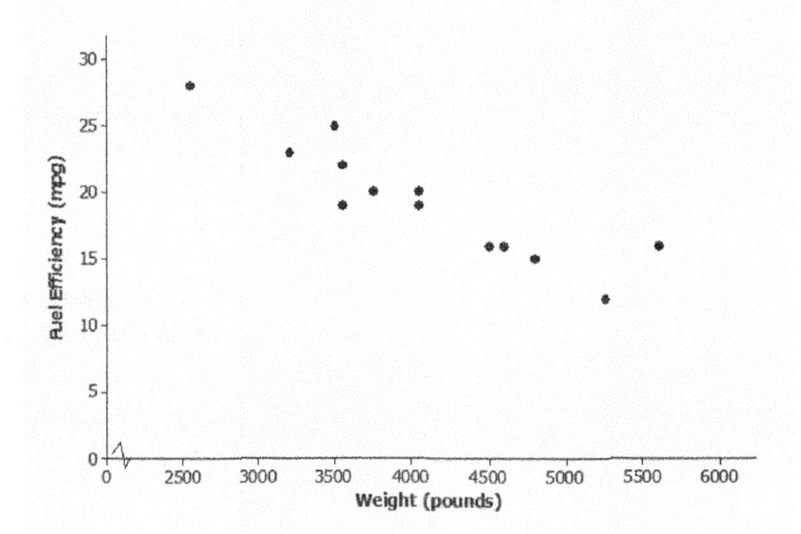

7. Below is a scatter plot of data on price in dollars (x) and quality rating (y) for 14 bike helmets. Using the questions at the beginning of this lesson as a guide, write a few sentences describing any possible relationship between x and y.

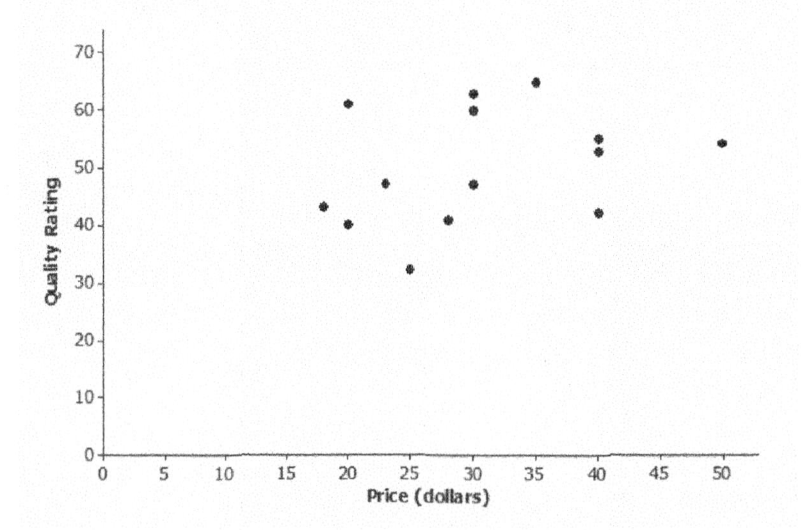

8. Below is a scatter plot of data on shell length in millimeters (x) and age in years (y) for 27 lobsters of known age. Using the questions at the beginning of this lesson as a guide, write a few sentences describing any possible relationship between x and y.

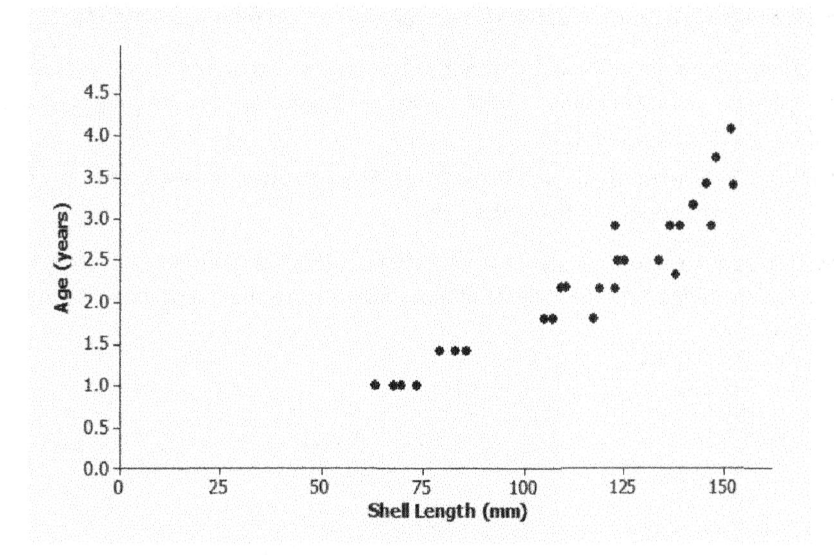

9. Below is a scatter plot of data from crocodiles on body mass in pounds (x) and bite force in pounds (y). Using the questions at the beginning of this lesson as a guide, write a few sentences describing any possible relationship between x and y.

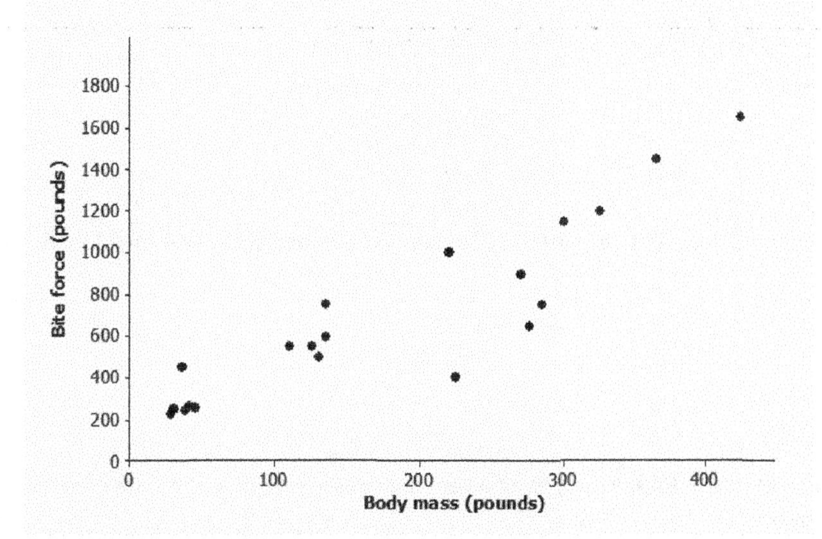

Data Source: http://journals.plos.org/plosone/article?id=10.1371/journal.pone.0031781#pone-0031781-t001

(Note: Body mass and bite force have been converted to pounds from kilograms and newtons, respectively.)

Example 2: Clusters and Outliers

In addition to looking for a general pattern in a scatter plot, you should also look for other interesting features that might help you understand the relationship between two variables. Two things to watch for are as follows:

- **CLUSTERS:** Usually, the points in a scatter plot form a single cloud of points, but sometimes the points may form two or more distinct clouds of points. These clouds are called *clusters*. Investigating these clusters may tell you something useful about the data.
- **OUTLIERS:** An *outlier* is an unusual point in a scatter plot that does not seem to fit the general pattern or that is far away from the other points in the scatter plot.

The scatter plot below was constructed using data from a study of Rocky Mountain elk ("Estimating Elk Weight from Chest Girth," *Wildlife Society Bulletin*, 1996). The variables studied were chest girth in centimeters (x) and weight in kilograms (y).

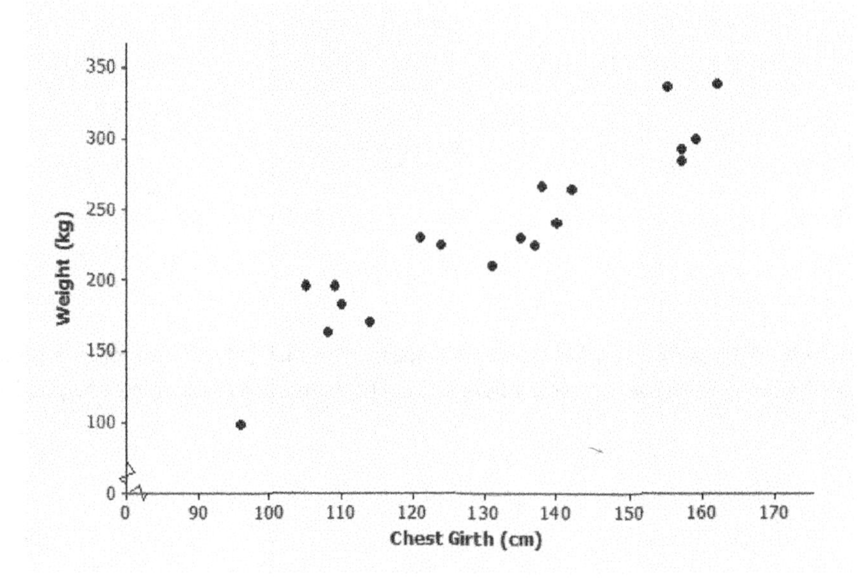

Exercises 10–12

10. Do you notice any point in the scatter plot of elk weight versus chest girth that might be described as an outlier? If so, which one?

11. If you identified an outlier in Exercise 10, write a sentence describing how this data observation differs from the others in the data set.

12. Do you notice any clusters in the scatter plot? If so, how would you distinguish between the clusters in terms of chest girth? Can you think of a reason these clusters might have occurred?

Lesson Summary

- A scatter plot might show a linear relationship, a nonlinear relationship, or no relationship.
- A positive linear relationship is one that would be modeled using a line with a positive slope. A negative linear relationship is one that would be modeled by a line with a negative slope.
- Outliers in a scatter plot are unusual points that do not seem to fit the general pattern in the plot or that are far away from the other points in the scatter plot.
- Clusters occur when the points in the scatter plot appear to form two or more distinct clouds of points.

Problem Set

1. Suppose data was collected on size in square feet (x) of several houses and price in dollars (y). The data was then used to construct the scatterplot below. Write a few sentences describing the relationship between price and size for these houses. Are there any noticeable clusters or outliers?

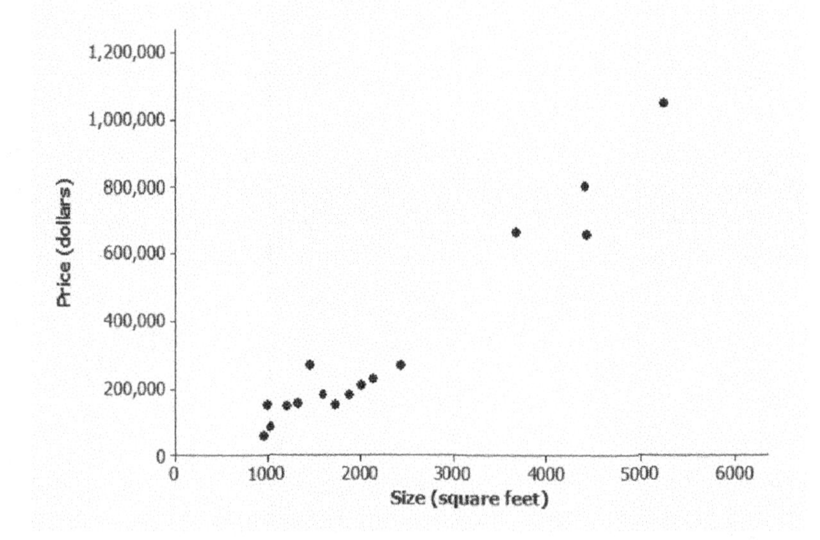

2. The scatter plot below was constructed using data on length in inches (x) of several alligators and weight in pounds (y). Write a few sentences describing the relationship between weight and length for these alligators. Are there any noticeable clusters or outliers?

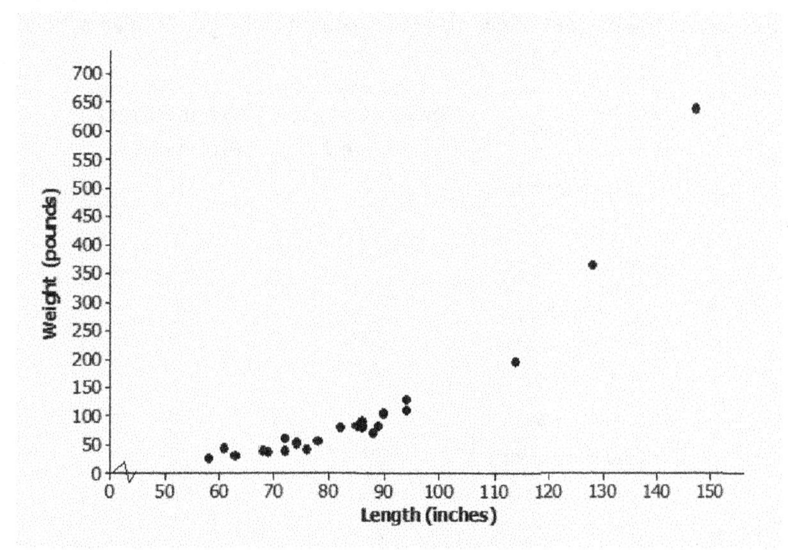

Data Source: Exploring Data, Quantitative Literacy Series, James Landwehr and Ann Watkins, 1987.

3. Suppose the scatter plot below was constructed using data on age in years (x) of several Honda Civics and price in dollars (y). Write a few sentences describing the relationship between price and age for these cars. Are there any noticeable clusters or outliers?

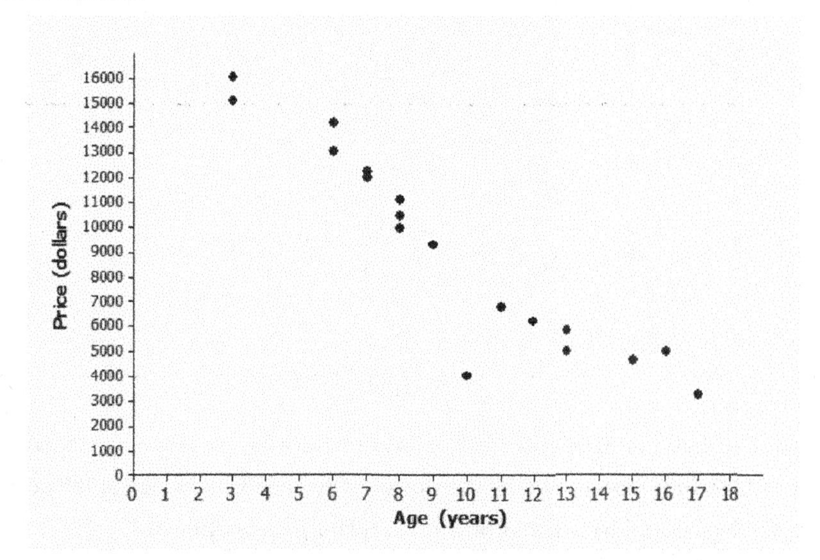

4. Samples of students in each of the U.S. states periodically take part in a large-scale assessment called the National Assessment of Educational Progress (NAEP). The table below shows the percent of students in the northeastern states (as defined by the U.S. Census Bureau) who answered Problems 7 and 15 correctly on the 2011 eighth-grade test. The scatter plot shows the percent of eighth-grade students who got Problems 7 and 15 correct on the 2011 NAEP.

State	Percent Correct Problem 7	Percent Correct Problem 15
Connecticut	29	51
New York	28	47
Rhode Island	29	52
Maine	27	50
Pennsylvania	29	48
Vermont	32	58
New Jersey	35	54
New Hampshire	29	52
Massachusetts	35	56

Percent Correct for Problems 7 and 15 on 2011 Eighth-Grade NAEP

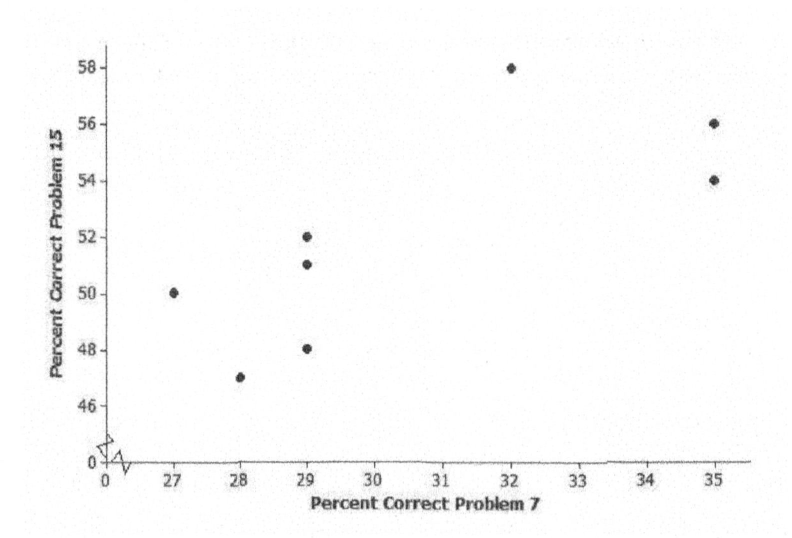

a. Why does it appear that there are only eight points in the scatter plot for nine states?

b. What is true of the states represented by the cluster of five points in the lower left corner of the graph?

c. Which state did the best on these two problems? Explain your reasoning.

d. Is there a trend in the data? Explain your thinking.

5. The plot below shows the mean percent of sunshine during the year and the mean amount of precipitation in inches per year for the states in the United States.

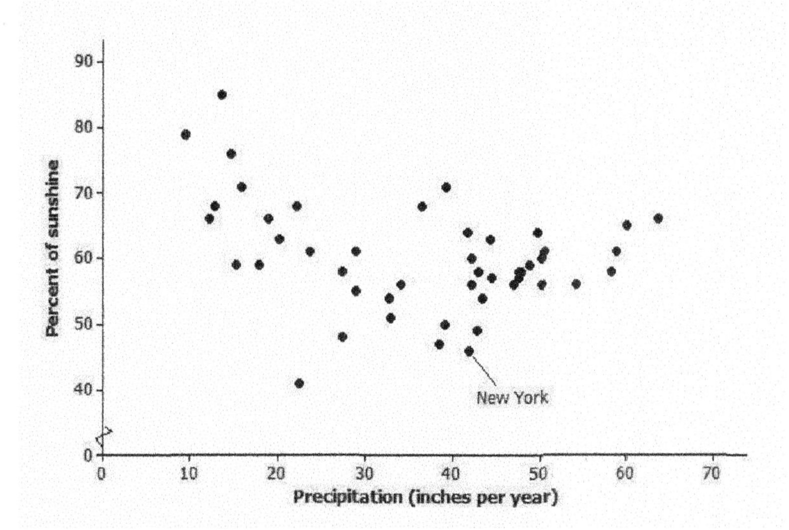

Data source: www.currentresults.com/Weather/US/average-annual-state-sunshine.php
www.currentresults.com/Weather/US/average-annual-state-precipitation.php

a. Where on the graph are the states that have a large amount of precipitation and a small percent of sunshine?

b. The state of New York is the point $(46, 41.8)$. Describe how the mean amount of precipitation and percent of sunshine in New York compare to the rest of the United States.

c. Write a few sentences describing the relationship between mean amount of precipitation and percent of sunshine.

6. At a dinner party, every person shakes hands with every other person present.

a. If three people are in a room and everyone shakes hands with everyone else, how many handshakes take place?

b. Make a table for the number of handshakes in the room for one to six people. You may want to make a diagram or list to help you count the number of handshakes.

Number People	Handshakes

Number People	Handshakes

c. Make a scatter plot of number of people (x) and number of handshakes (y). Explain your thinking.

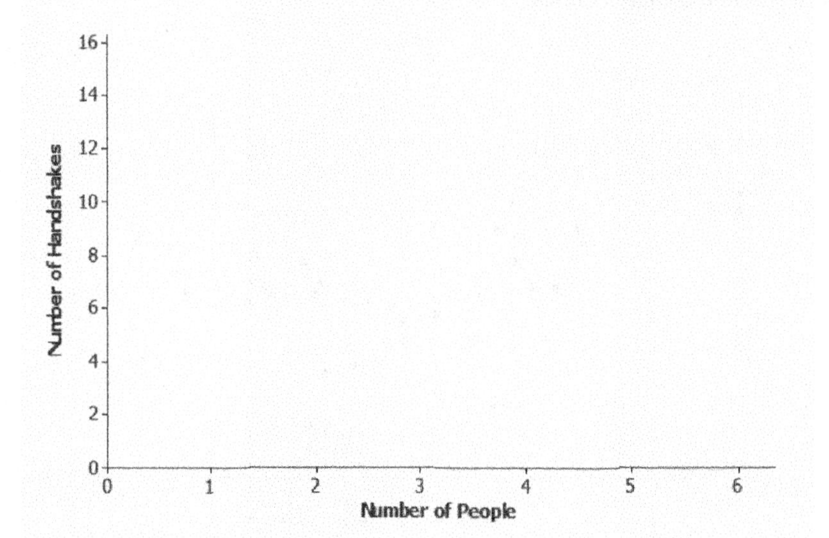

d. Does the trend seem to be linear? Why or why not?

Lesson 8: Informally Fitting a Line

Classwork

Example 1: Housing Costs

Let's look at some data from one midwestern city that indicate the sizes and sale prices of various houses sold in this city.

Size (square feet)	Price (dollars)
5,232	1,050,000
1,875	179,900
1,031	84,900
1,437	269,900
4,400	799,900
2,000	209,900
2,132	224,900
1,591	179,900

Size (square feet)	Price (dollars)
1,196	144,900
1,719	149,900
956	59,900
991	149,900
1,312	154,900
4,417	659,999
3,664	669,000
2,421	269,900

Data Source: http://www.trulia.com/for_sale/Milwaukee,WI/5_p, accessed in 2013

A scatter plot of the data is given below.

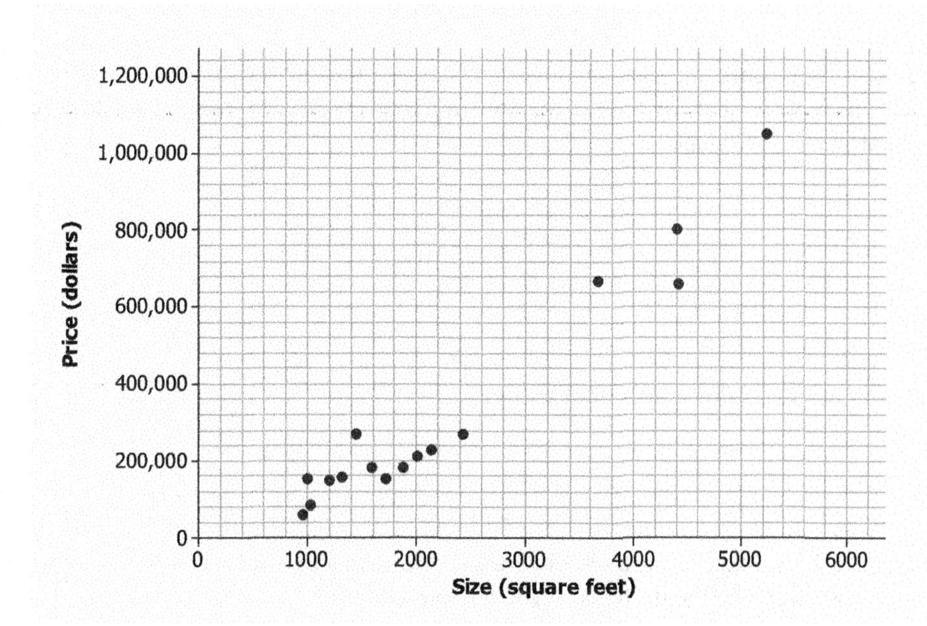

Exercises 1–6

1. What can you tell about the price of large homes compared to the price of small homes from the table?

2. Use the scatter plot to answer the following questions.

   a. Does the scatter plot seem to support the statement that larger houses tend to cost more? Explain your thinking.

   b. What is the cost of the most expensive house, and where is that point on the scatter plot?

   c. Some people might consider a given amount of money and then predict what size house they could buy. Others might consider what size house they want and then predict how much it would cost. How would you use the scatter plot in Example 1?

   d. Estimate the cost of a $3{,}000$-square-foot house.

   e. Do you think a line would provide a reasonable way to describe how price and size are related? How could you use a line to predict the price of a house if you are given its size?

3. Draw a line in the plot that you think would fit the trend in the data.

4. Use your line to answer the following questions:

   a. What is your prediction of the price of a $3{,}000$-square-foot house?

   b. What is the prediction of the price of a $1{,}500$-square-foot house?

5. Consider the following general strategies students use for drawing a line. Do you think they represent a good strategy for drawing a line that fits the data? Explain why or why not, or draw a line for the scatter plot using the strategy that would indicate why it is or why it is not a good strategy.

   a. Laure thought she might draw her line using the very first point (farthest to the left) and the very last point (farthest to the right) in the scatter plot.

   b. Phil wants to be sure that he has the same number of points above and below the line.

   c. Sandie thought she might try to get a line that had the most points right on it.

   d. Maree decided to get her line as close to as many of the points as possible.

6. Based on the strategies discussed in Exercise 5, would you change how you draw a line through the points? Explain your answer.

Example 2: Deep Water

Does the current in the water go faster or slower when the water is shallow? The data on the depth and velocity of the Columbia River at various locations in Washington State listed below can help you think about the answer.

Depth and Velocity in the Columbia River, Washington State

Depth (feet)	Velocity (feet/second)
0.7	1.55
2.0	1.11
2.6	1.42
3.3	1.39
4.6	1.39
5.9	1.14
7.3	0.91
8.6	0.59
9.9	0.59
10.6	0.41
11.2	0.22

Data Source: www.seattlecentral.edu/qelp/sets/011/011.html

a. What can you tell about the relationship between the depth and velocity by looking at the numbers in the table?

b. If you were to make a scatter plot of the data, which variable would you put on the horizontal axis, and why?

Exercises 7–9

7. A scatter plot of the Columbia River data is shown below.

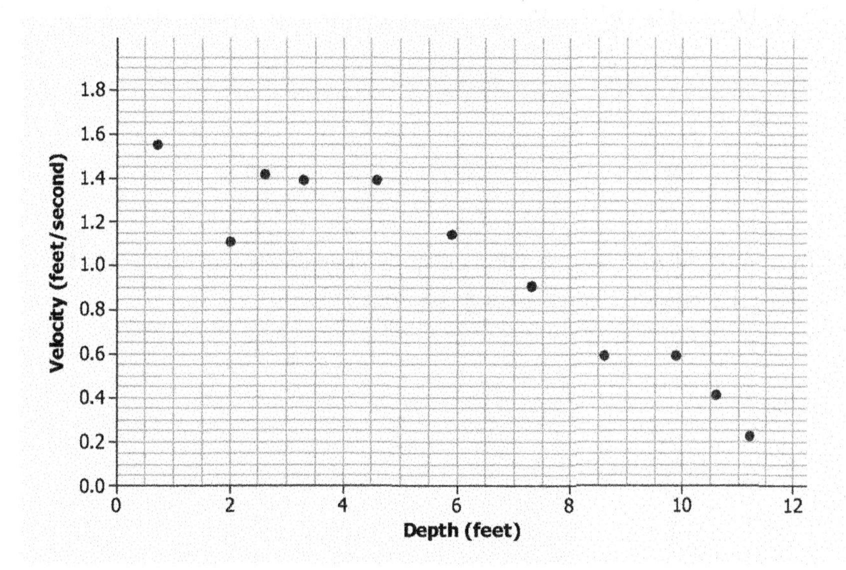

a. Choose a data point in the scatter plot, and describe what it means in terms of the context.

b. Based on the scatter plot, describe the relationship between velocity and depth.

c. How would you explain the relationship between the velocity and depth of the water?

d. If the river is two feet deep at a certain spot, how fast do you think the current would be? Explain your reasoning.

8. Consider the following questions:

   a. If you draw a line to represent the trend in the plot, will it make it easier to predict the velocity of the water if you know the depth? Why or why not?

   b. Draw a line that you think does a reasonable job of modeling the trend on the scatter plot in Exercise 7. Use the line to predict the velocity when the water is 8 feet deep.

9. Use the line to predict the velocity for a depth of 8.6 feet. How far off was your prediction from the actual observed velocity for the location that had a depth of 8.6 feet?

Lesson Summary

- When constructing a scatter plot, the variable that you want to predict (i.e., the dependent or response variable) goes on the vertical axis. The independent variable (i.e., the variable not related to other variables) goes on the horizontal axis.
- When the pattern in a scatter plot is approximately linear, a line can be used to describe the linear relationship.
- A line that describes the relationship between a dependent variable and an independent variable can be used to make predictions of the value of the dependent variable given a value of the independent variable.
- When informally fitting a line, you want to find a line for which the points in the scatter plot tend to be closest.

Problem Set

1. The table below shows the mean temperature in July and the mean amount of rainfall per year for 14 cities in the Midwest.

City	Mean Temperature in July (degrees Fahrenheit)	Mean Rainfall per Year (inches)
Chicago, IL	73.3	36.27
Cleveland, OH	71.9	38.71
Columbus, OH	75.1	38.52
Des Moines, IA	76.1	34.72
Detroit, MI	73.5	32.89
Duluth, MN	65.5	31.00
Grand Rapids, MI	71.4	37.13
Indianapolis, IN	75.4	40.95
Marquette, MI	71.6	32.95
Milwaukee, WI	72.0	34.81
Minneapolis–St. Paul, MN	73.2	29.41
Springfield, MO	76.3	35.56
St. Louis, MO	80.2	38.75
Rapid City, SD	73.0	33.21

Data Source: http://countrystudies.us/united-states/weather/

a. What do you observe from looking at the data in the table?

b. Look at the scatter plot below. A line is drawn to fit the data. The plot in the Exit Ticket had the mean July temperatures for the cities on the horizontal axis. How is this plot different, and what does it mean for the way you think about the relationship between the two variables—temperature and rain?

July Rainfall and Temperatures in Selected Midwestern Cities

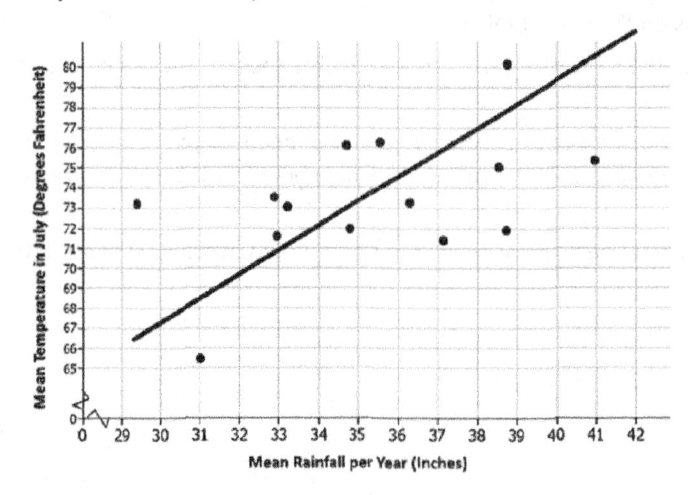

c. The line has been drawn to model the relationship between the amount of rain and the temperature in those midwestern cities. Use the line to predict the mean July temperature for a midwestern city that has a mean of 32 inches of rain per year.

d. For which of the cities in the sample does the line do the worst job of predicting the mean temperature? The best? Explain your reasoning with as much detail as possible.

2. The scatter plot below shows the results of a survey of eighth-grade students who were asked to report the number of hours per week they spend playing video games and the typical number of hours they sleep each night.

Mean Hours Sleep per Night Versus Mean Hours Playing Video Games per Week

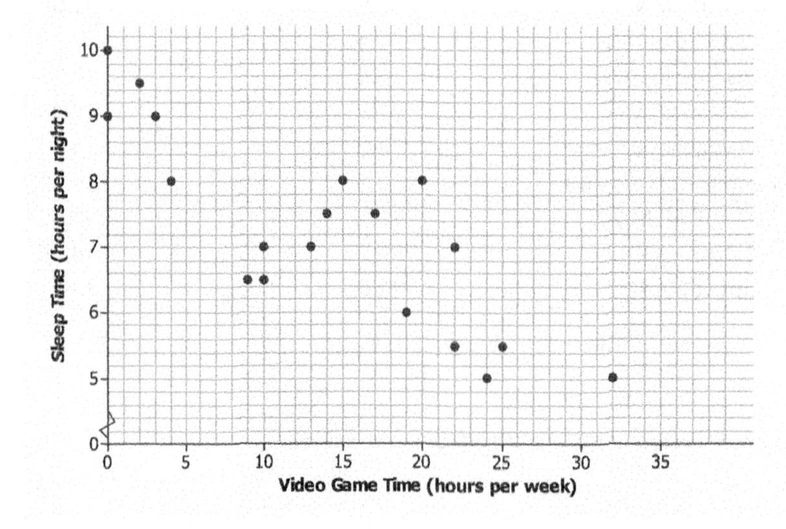

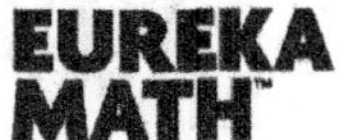

a. What trend do you observe in the data?

b. What was the fewest number of hours per week that students who were surveyed spent playing video games? The most?

c. What was the fewest number of hours per night that students who were surveyed typically slept? The most?

d. Draw a line that seems to fit the trend in the data, and find its equation. Use the line to predict the number of hours of sleep for a student who spends about 15 hours per week playing video games.

3. Scientists can take very good pictures of alligators from airplanes or helicopters. Scientists in Florida are interested in studying the relationship between the length and the weight of alligators in the waters around Florida.

   a. Would it be easier to collect data on length or weight? Explain your thinking.

   b. Use your answer to decide which variable you would want to put on the horizontal axis and which variable you might want to predict.

4. Scientists captured a small sample of alligators and measured both their length (in inches) and weight (in pounds). Torre used their data to create the following scatter plot and drew a line to capture the trend in the data. She and Steve then had a discussion about the way the line fit the data. What do you think they were discussing, and why?

Alligator Length (inches) and Weight (pounds)

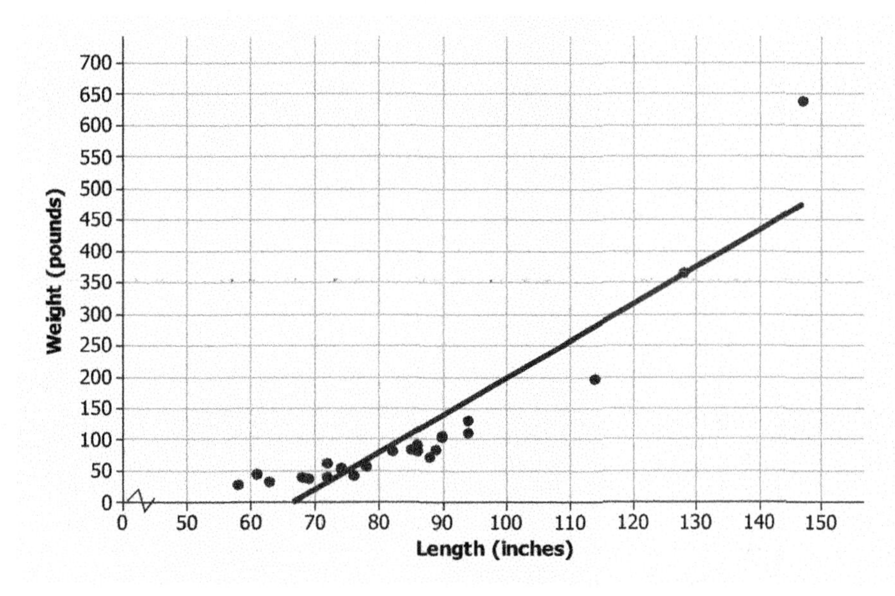

Data Source: James Landwehr and Ann Watkins, *Exploring Data*, Quantitative Literacy Series (Dale Seymour, 1987).

This page intentionally left blank

Lesson 9: Determining the Equation of a Line Fit to Data

Classwork

Example 1: Crocodiles and Alligators

Scientists are interested in finding out how different species adapt to finding food sources. One group studied crocodilians to find out how their bite force was related to body mass and diet. The table below displays the information they collected on body mass (in pounds) and bite force (in pounds).

Crocodilian Biting

Species	Body Mass (pounds)	Bite Force (pounds)
Dwarf crocodile	35	450
Crocodile F	40	260
Alligator A	30	250
Caiman A	28	230
Caiman B	37	240
Caiman C	45	255
Crocodile A	110	550
Nile crocodile	275	650
Crocodile B	130	500
Crocodile C	135	600
Crocodile D	135	750
Caiman D	125	550
Indian gharial crocodile	225	400
Crocodile G	220	1,000
American crocodile	270	900
Crocodile E	285	750
Crocodile F	425	1,650
American alligator	300	1,150
Alligator B	325	1,200
Alligator C	365	1,450

Data Source: http://journals.plos.org/plosone/article?id=10.1371/journal.pone.0031781#pone-0031781-t001

(Note: Body mass and bite force have been converted to pounds from kilograms and newtons, respectively.)

As you learned in the previous lesson, it is a good idea to begin by looking at what a scatter plot tells you about the data. The scatter plot below displays the data on body mass and bite force for the crocodilians in the study.

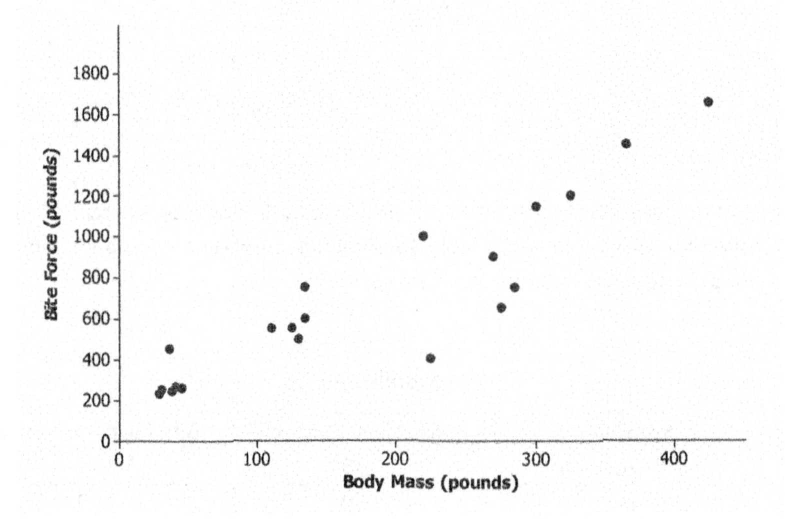

Exercises 1–6

1. Describe the relationship between body mass and bite force for the crocodilians shown in the scatter plot.

2. Draw a line to represent the trend in the data. Comment on what you considered in drawing your line.

3. Based on your line, predict the bite force for a crocodilian that weighs 220 pounds. How does this prediction compare to the actual bite force of the 220-pound crocodilian in the data set?

4. Several students decided to draw lines to represent the trend in the data. Consider the lines drawn by Sol, Patti, Marrisa, and Taylor, which are shown below.

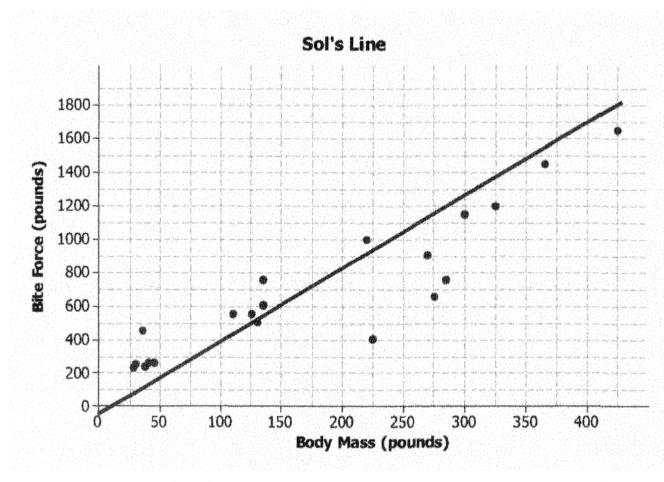

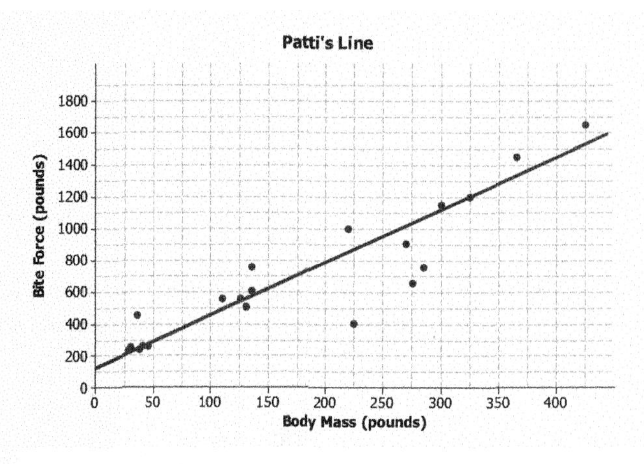

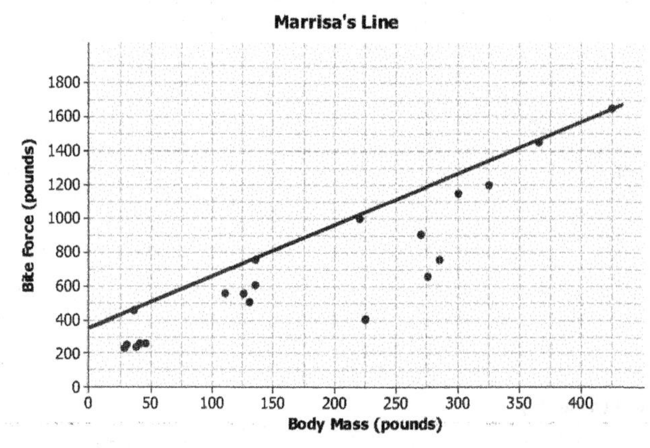

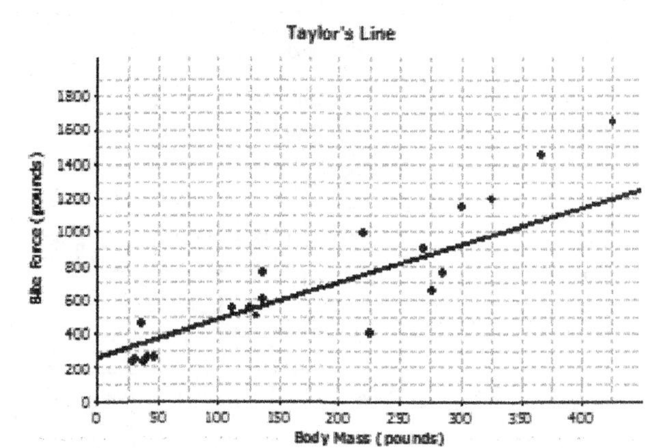

For each student, indicate whether or not you think the line would be a good line to use to make predictions. Explain your thinking.

a. Sol's line

b. Patti's line

c. Marrisa's line

d. Taylor's line

5. What is the equation of your line? Show the steps you used to determine your line. Based on your equation, what is your prediction for the bite force of a crocodilian weighing 200 pounds?

6. Patti drew vertical line segments from two points to the line in her scatter plot. The first point she selected was for a dwarf crocodile. The second point she selected was for an Indian gharial crocodile.

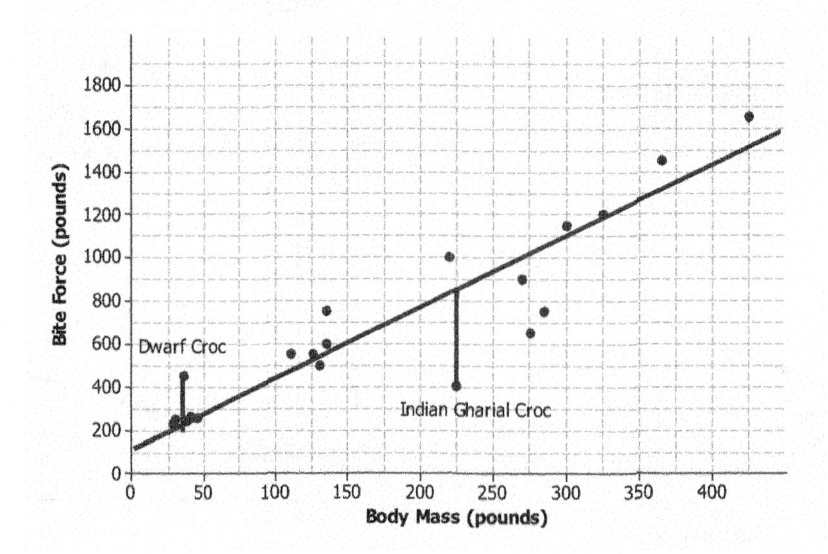

a. Would Patti's line have resulted in a predicted bite force that was closer to the actual bite force for the dwarf crocodile or for the Indian gharial crocodile? What aspect of the scatter plot supports your answer?

b. Would it be preferable to describe the trend in a scatter plot using a line that makes the differences in the actual and predicted values large or small? Explain your answer.

Exercise 7: Used Cars

7. Suppose the plot below shows the age (in years) and price (in dollars) of used compact cars that were advertised in a local newspaper.

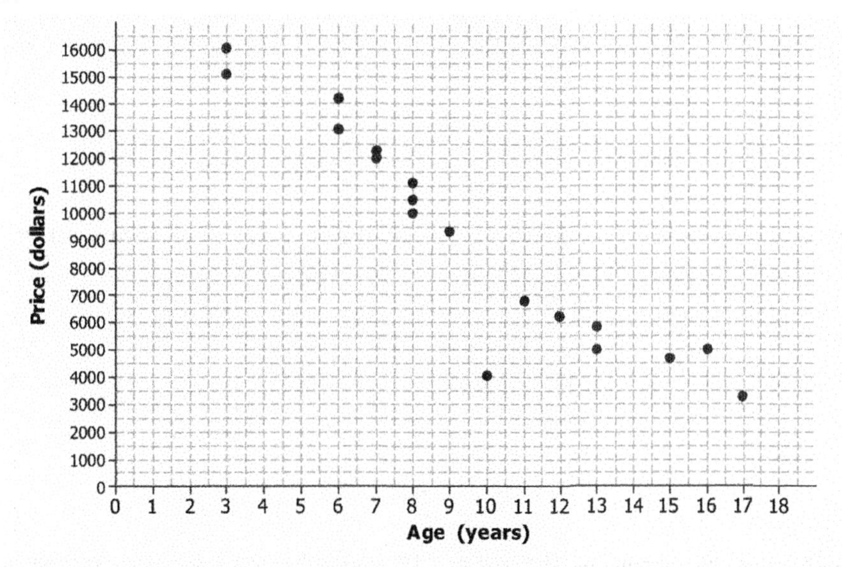

a. Based on the scatter plot above, describe the relationship between the age and price of the used cars.

b. Nora drew a line she thought was close to many of the points and found the equation of the line. She used the points $(13, 6000)$ and $(7, 12000)$ on her line to find the equation. Explain why those points made finding the equation easy.

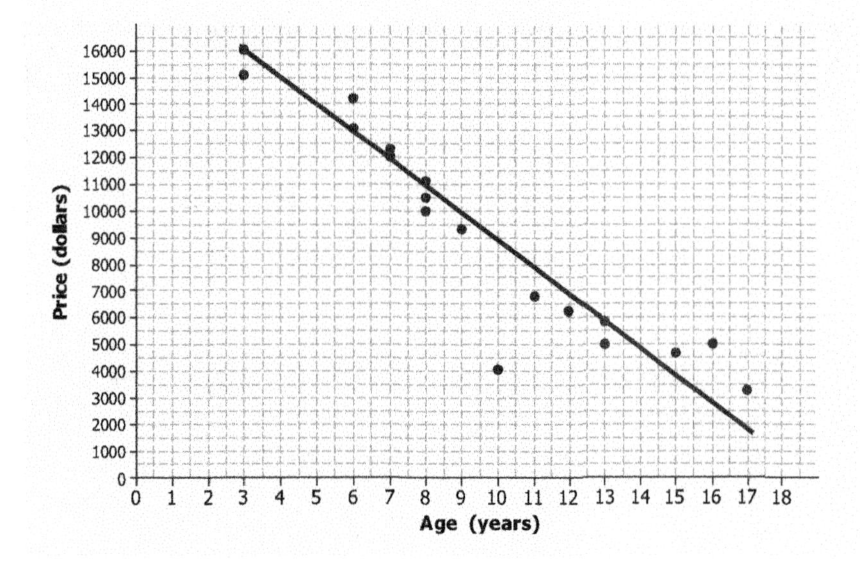

c. Find the equation of Nora's line for predicting the price of a used car given its age. Summarize the trend described by this equation.

d. Based on the line, for which car in the data set would the predicted value be farthest from the actual value? How can you tell?

e. What does the equation predict for the cost of a 10-year-old car? How close was the prediction using the line to the actual cost of the 10-year-old car in the data set? Given the context of the data set, do you think the difference between the predicted price and the actual price is large or small?

f. Is $5,000 typical of the differences between predicted prices and actual prices for the cars in this data set? Justify your answer.

Lesson Summary

- A line can be used to represent the trend in a scatter plot.
- Evaluating the equation of the line for a value of the independent variable determines a value predicted by the line.
- A good line for prediction is one that goes through the middle of the points in a scatter plot and for which the points tend to fall close to the line.

Problem Set

1. The Monopoly board game is popular in many countries. The scatter plot below shows the distance from "Go" to a property (in number of spaces moving from "Go" in a clockwise direction) and the price of the properties on the Monopoly board. The equation of the line is $P = 8x + 40$, where P represents the price (in Monopoly dollars) and x represents the distance (in number of spaces).

Distance from "Go" (number of spaces)	Price of Property (Monopoly dollars)
1	60
3	60
5	200
6	100
8	100
9	120
11	140
12	150
13	140
14	160
15	200
16	180
18	180
19	200

Distance from "Go" (number of spaces)	Price of Property (Monopoly dollars)
21	220
23	220
24	240
25	200
26	260
27	260
28	150
29	280
31	300
32	300
34	320
35	200
37	350
39	400

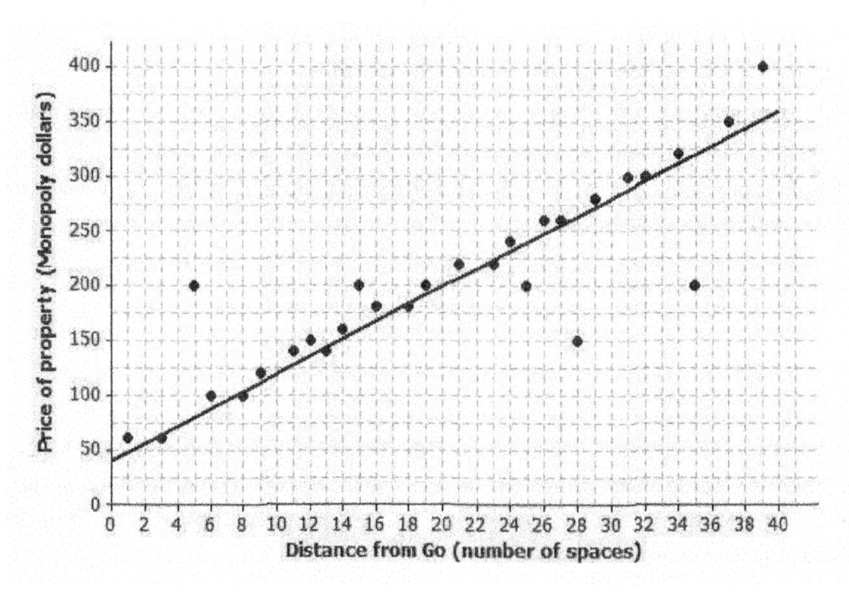

a. Use the equation to find the difference (observed value−predicted value) for the most expensive property and for the property that is 35 spaces from "Go."

b. Five of the points seem to lie in a horizontal line. What do these points have in common? What is the equation of the line containing those five points?

c. Four of the five points described in part (b) are the railroads. If you were fitting a line to predict price with distance from "Go," would you use those four points? Why or why not?

2. The table below gives the coordinates of the five points shown in the scatter plots that follow. The scatter plots show two different lines.

Data Point	Independent Variable	Response Variable
A	20	27
B	22	21
C	25	24
D	31	18
E	40	12

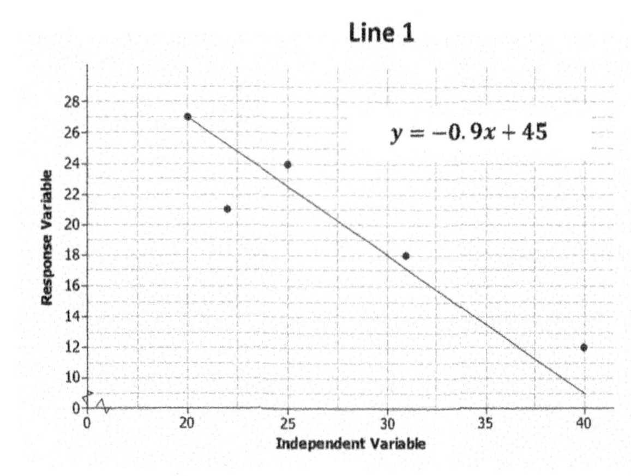

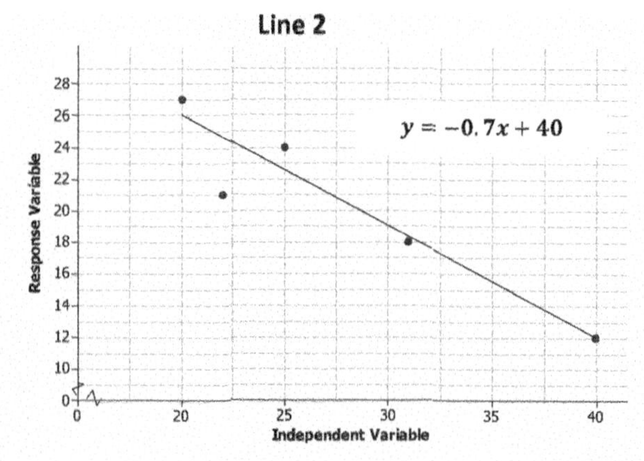

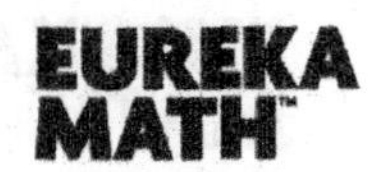

a. Find the predicted response values for each of the two lines.

Independent	Observed Response	Response Predicted by Line 1	Response Predicted by Line 2

b. For which data points is the prediction based on Line 1 closer to the actual value than the prediction based on Line 2?

c. Which line (Line 1 or Line 2) would you select as a better fit? Explain.

3. The scatter plots below show different lines that students used to model the relationship between body mass (in pounds) and bite force (in pounds) for crocodilians.

   a. Match each graph to one of the equations below, and explain your reasoning. Let B represent bite force (in pounds) and W represent body mass (in pounds).

Equation 1	Equation 2	Equation 3
$B = 3.28W + 126$	$B = 3.04W + 351$	$B = 2.16W + 267$

Equation:

Equation:

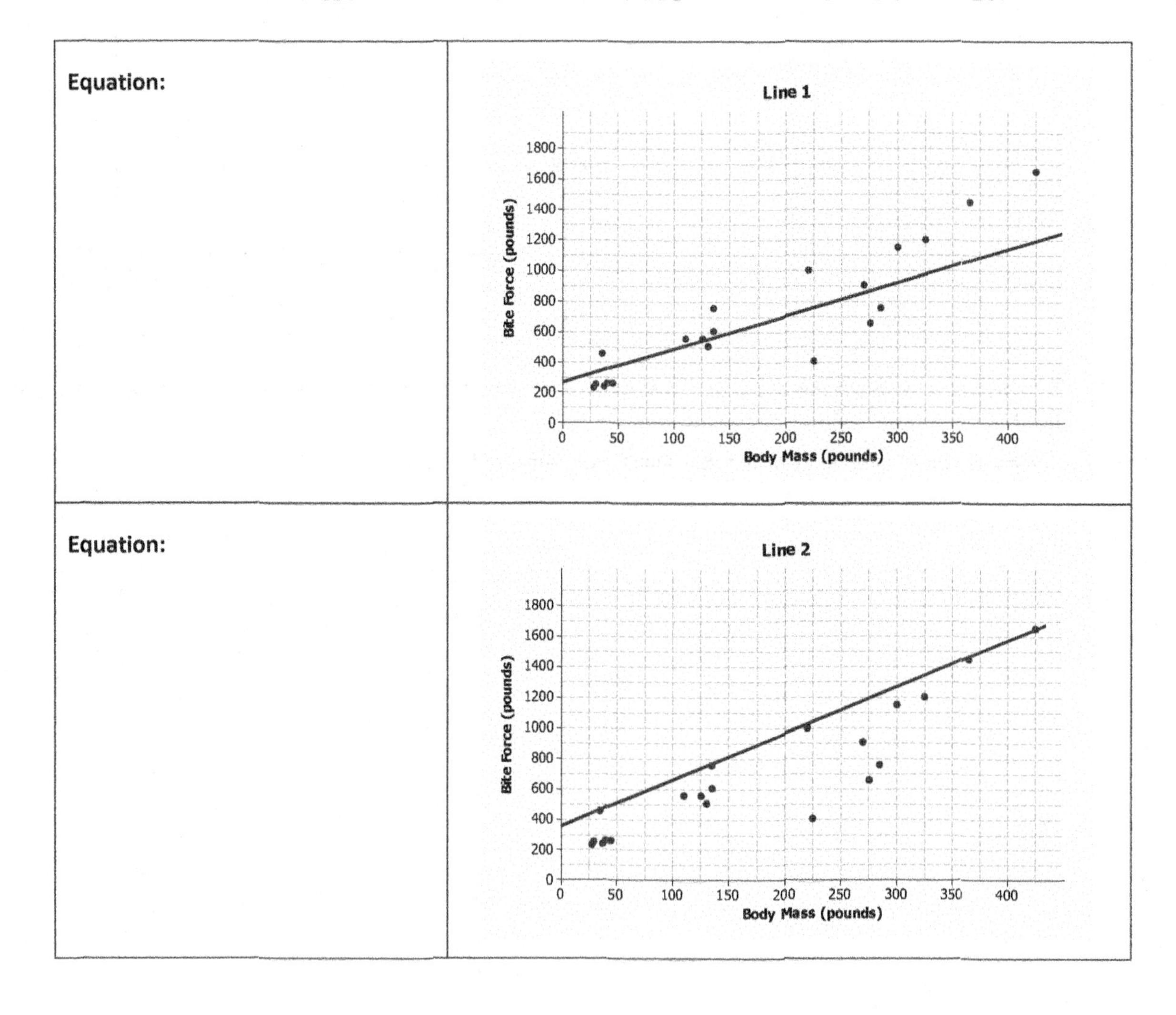

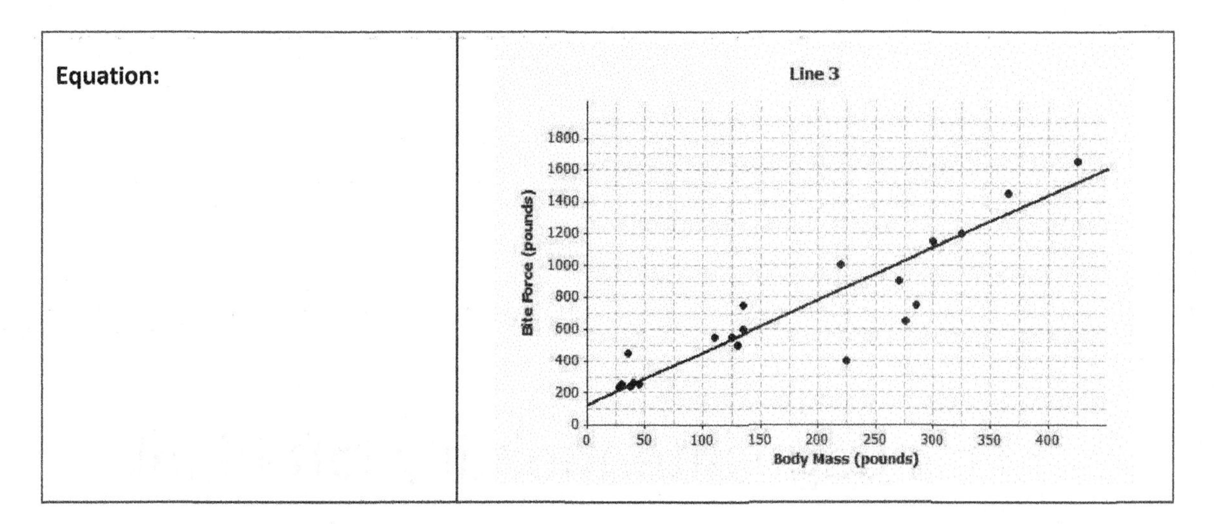

b. Which of the lines would best fit the trend in the data? Explain your thinking.

4. Comment on the following statements:

   a. A line modeling a trend in a scatter plot always goes through the origin.

   b. If the response variable increases as the independent variable decreases, the slope of a line modeling the trend is negative.

This page intentionally left blank

Lesson 10: Linear Models

Classwork

In previous lessons, you used data that follow a linear trend either in the positive direction or the negative direction and informally fit a line through the data. You determined the equation of an informal fitted line and used it to make predictions.

In this lesson, you use a function to model a linear relationship between two numerical variables and interpret the slope and intercept of the linear model in the context of the data. Recall that a function is a rule that relates a dependent variable to an independent variable.

In statistics, a dependent variable is also called a *response variable* or a *predicted variable*. An independent variable is also called an *explanatory variable* or a *predictor variable*.

Example 1

Predicting the value of a numerical dependent (response) variable based on the value of a given numerical independent variable has many applications in statistics. The first step in the process is to identify the dependent (predicted) variable and the independent (predictor) variable.

There may be several independent variables that might be used to predict a given dependent variable. For example, suppose you want to predict how well you are going to do on an upcoming statistics quiz. One possible independent variable is how much time you spent studying for the quiz. What are some other possible numerical independent variables that could relate to how well you are going to do on the quiz?

Exercises 1–2

1. For each of the following dependent (response) variables, identify two possible numerical independent (explanatory) variables that might be used to predict the value of the dependent variable.

Response Variable	Possible Explanatory Variables
Height of a son	
Number of points scored in a game by a basketball player	
Number of hamburgers to make for a family picnic	
Time it takes a person to run a mile	
Amount of money won by a contestant on Jeopardy!™ (television game show)	

Response Variable	Possible Explanatory Variables
Fuel efficiency (in miles per gallon) for a car	
Number of honey bees in a beehive at a particular time	
Number of blooms on a dahlia plant	
Number of forest fires in a state during a particular year	

2. Now, reverse your thinking. For each of the following numerical independent variables, write a possible numerical dependent variable.

Dependent Variable	Possible Independent Variables
	Age of a student
	Height of a golfer
	Amount of a pain reliever taken
	Number of years of education
	Amount of fertilizer used on a garden
	Size of a diamond in a ring
	Total salary for all of a team's players

Example 2

A cell phone company offers the following basic cell phone plan to its customers: A customer pays a monthly fee of $40.00. In addition, the customer pays $0.15 per text message sent from the cell phone. There is no limit to the number of text messages per month that could be sent, and there is no charge for receiving text messages.

Exercises 3–11

3. Determine the following:

   a. Justin never sends a text message. What would be his total monthly cost?

   b. During a typical month, Abbey sends 25 text messages. What is her total cost for a typical month?

   c. Robert sends at least 250 text messages a month. What would be an estimate of the least his total monthly cost is likely to be?

4. Use descriptive words to write a linear model describing the relationship between the number of text messages sent and the total monthly cost.

5. Is the relationship between the number of text messages sent and the total monthly cost linear? Explain your answer.

6. Let x represent the independent variable and y represent the dependent variable. Use the variables x and y to write the function representing the relationship you indicated in Exercise 4.

7. Explain what $\$0.15$ represents in this relationship.

8. Explain what $\$40.00$ represents in this relationship.

9. Sketch a graph of this relationship on the following coordinate grid. Clearly label the axes, and include units in the labels.

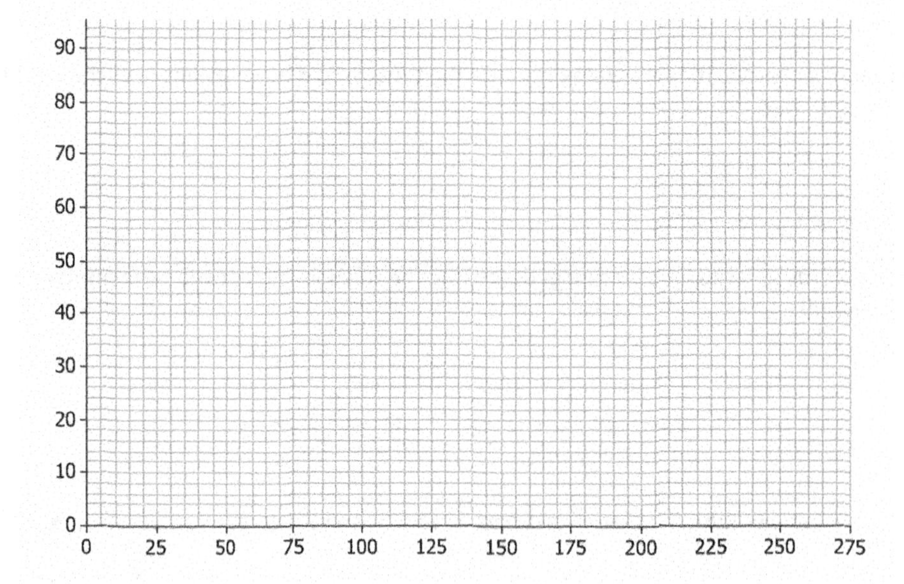

10. LaMoyne needs four more pieces of lumber for his Scout project. The pieces can be cut from one large piece of lumber according to the following pattern.

The lumberyard will make the cuts for LaMoyne at a fixed cost of $\$2.25$ plus an additional cost of 25 cents per cut. One cut is free.

a. What is the functional relationship between the total cost of cutting a piece of lumber and the number of cuts required? What is the equation of this function? Be sure to define the variables in the context of this problem.

b. Use the equation to determine LaMoyne's total cost for cutting.

c. In the context of this problem, interpret the slope of the equation in words.

d. Interpret the y-intercept of your equation in words in the context of this problem. Does interpreting the intercept make sense in this problem? Explain.

11. Omar and Olivia were curious about the size of coins. They measured the diameter and circumference of several coins and found the following data.

U.S. Coin	Diameter (millimeters)	Circumference (millimeters)
Penny	19.0	59.7
Nickel	21.2	66.6
Dime	17.9	56.2
Quarter	24.3	76.3
Half Dollar	30.6	96.1

a. Wondering if there was any relationship between diameter and circumference, they thought about drawing a picture. Draw a scatter plot that displays circumference in terms of diameter.

b. Do you think that circumference and diameter are related? Explain.

c. Find the equation of the function relating circumference to the diameter of a coin.

d. The value of the slope is approximately equal to the value of π. Explain why this makes sense.

e. What is the value of the y-intercept? Explain why this makes sense.

Lesson Summary

- A linear functional relationship between a dependent and an independent numerical variable has the form $y = mx + b$ or $y = a + bx$.
- In statistics, a dependent variable is one that is predicted, and an independent variable is the one that is used to make the prediction.
- The graph of a linear function describing the relationship between two variables is a line.

Problem Set

1. The Mathematics Club at your school is having a meeting. The advisor decides to bring bagels and his award-winning strawberry cream cheese. To determine his cost, from past experience he figures 1.5 bagels per student. A bagel costs 65 cents, and the special cream cheese costs \$3.85 and will be able to serve all of the anticipated students attending the meeting
   a. Find an equation that relates his total cost to the number of students he thinks will attend the meeting.
   b. In the context of the problem, interpret the slope of the equation in words.
   c. In the context of the problem, interpret the y-intercept of the equation in words. Does interpreting the intercept make sense? Explain.

2. John, Dawn, and Ron agree to walk/jog for 45 minutes. John has arthritic knees but manages to walk $1\frac{1}{2}$ miles. Dawn walks $2\frac{1}{4}$ miles, while Ron manages to jog 6 miles.
   a. Draw an appropriate graph, and connect the points to show that there is a linear relationship between the distance that each traveled based on how fast each traveled (speed). Note that the speed for a person who travels 3 miles in 45 minutes, or $\frac{3}{4}$ hour, is found using the expression $3 \div \frac{3}{4}$, which is 4 miles per hour.
   b. Find an equation that expresses distance in terms of speed (how fast one goes).
   c. In the context of the problem, interpret the slope of the equation in words.
   d. In the context of the problem, interpret the y-intercept of the equation in words. Does interpreting the intercept make sense? Explain.

3. Simple interest is money that is paid on a loan. Simple interest is calculated by taking the amount of the loan and multiplying it by the rate of interest per year and the number of years the loan is outstanding. For college, Jodie's older brother has taken out a student loan for $\$4{,}500$ at an annual interest rate of 5.6%, or 0.056. When he graduates in four years, he has to pay back the loan amount plus interest for four years. Jodie is curious as to how much her brother has to pay.

   a. Jodie claims that her brother has to pay a total of $\$5{,}508$. Do you agree? Explain. As an example, a $\$1{,}200$ loan has an 8% annual interest rate. The simple interest for one year is $\$96$ because $(0.08)(1200) = 96$. The interest for two years would be $\$192$ because $(2)(96) = 192$.

   b. Write an equation for the total cost to repay a loan of $\$P$ if the rate of interest for a year is r (expressed as a decimal) for a time span of t years.

   c. If P and r are known, is the equation a linear equation?

   d. In the context of the problem, interpret the slope of the equation in words.

   e. In the context of the problem, interpret the y-intercept of the equation in words. Does interpreting the intercept make sense? Explain.

This page intentionally left blank

Lesson 11: Using Linear Models in a Data Context

Classwork

Exercises

1. Old Faithful is a geyser in Yellowstone National Park. The following table offers some rough estimates of the length of an eruption (in minutes) and the amount of water (in gallons) in that eruption.

Length (minutes)	1.5	2	3	4.5
Amount of Water (gallons)	3,700	4,100	6,450	8,400

This data is consistent with actual eruption and summary statistics that can be found at the following links:

http://geysertimes.org/geyser.php?id=OldFaithful and http://www.yellowstonepark.com/2011/07/about-old-faithful/

a. Chang wants to predict the amount of water in an eruption based on the length of the eruption. What should he use as the dependent variable? Why?

b. Which of the following two scatter plots should Chang use to build his prediction model? Explain.

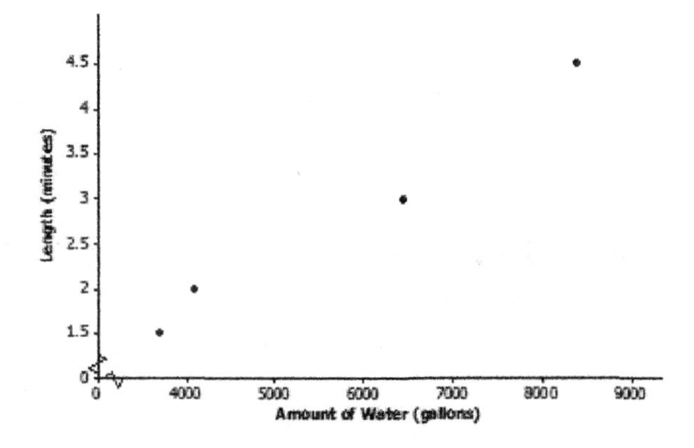

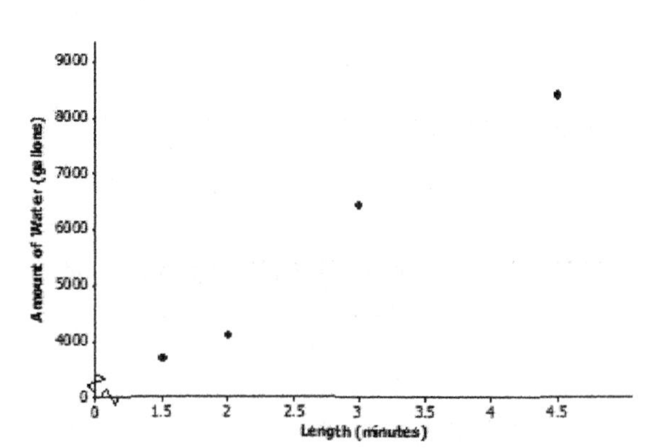

c. Suppose that Chang believes the variables to be linearly related. Use the *first* and *last* data points in the table to create a linear prediction model.

d. A friend of Chang's told him that Old Faithful produces about 3,000 gallons of water for every minute that it erupts. Does the linear model from part (c) support what Chang's friend said? Explain.

e. Using the linear model from part (c), does it make sense to interpret the y-intercept in the context of this problem? Explain.

2. The following table gives the times of the gold, silver, and bronze medal winners for the men's 100-meter race (in seconds) for the past 10 Olympic Games.

Year	2012	2008	2004	2000	1996	1992	1988	1984	1980	1976
Gold	9.63	9.69	9.85	9.87	9.84	9.96	9.92	9.99	10.25	10.06
Silver	9.75	9.89	9.86	9.99	9.89	10.02	9.97	10.19	10.25	10.07
Bronze	9.79	9.91	9.87	10.04	9.90	10.04	9.99	10.22	10.39	10.14
Mean Time	9.72	9.83	9.86	9.97	9.88	10.01	9.96	10.13	10.30	10.09

Data Source: https://en.wikipedia.org/wiki/100_metres_at_the_Olympics#Men

a. If you wanted to describe how mean times change over the years, which variable would you use as the independent variable, and which would you use as the dependent variable?

b. Draw a scatter plot to determine if the relationship between mean time and year appears to be linear. Comment on any trend or pattern that you see in the scatter plot.

c. One reasonable line goes through the 1992 and 2004 data. Find the equation of that line.

d. Before he saw these data, Chang guessed that the mean time of the three Olympic medal winners decreased by about 0.05 seconds from one Olympic Game to the next. Does the prediction model you found in part (c) support his guess? Explain.

e. If the trend continues, what mean race time would you predict for the gold, silver, and bronze medal winners in the 2016 Olympic Games? Explain how you got this prediction.

f. The data point $(1980, 10.3)$ appears to have an unusually high value for the mean race time (10.3). Using your library or the Internet, see if you can find a possible explanation for why that might have happened.

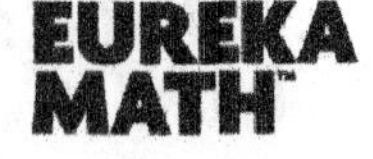

Lesson Summary

In the real world, it is rare that two numerical variables are exactly linearly related. If the data are roughly linearly related, then a line can be drawn through the data. This line can then be used to make predictions and to answer questions. For now, the line is informally drawn, but in later grades more formal methods for determining a best-fitting line are presented.

Problem Set

1. From the United States Bureau of Census website, the population sizes (in millions of people) in the United States for census years 1790–2010 are as follows.

Year	1790	1800	1810	1820	1830	1840	1850	1860	1870	1880	1890
Population Size	3.9	5.3	7.2	9.6	12.9	17.1	23.2	31.4	38.6	50.2	63.0

Year	1900	1910	1920	1930	1940	1950	1960	1970	1980	1990	2000	2010
Population Size	76.2	92.2	106.0	123.2	132.2	151.3	179.3	203.3	226.5	248.7	281.4	308.7

   a. If you wanted to be able to predict population size in a given year, which variable would be the independent variable, and which would be the dependent variable?

   b. Draw a scatter plot. Does the relationship between year and population size appear to be linear?

   c. Consider the data only from 1950 to 2010. Does the relationship between year and population size for these years appear to be linear?

   d. One line that could be used to model the relationship between year and population size for the data from 1950 to 2010 is $y = -4875.021 + 2.578x$. Suppose that a sociologist believes that there will be negative consequences if population size in the United States increases by more than $2\frac{3}{4}$ million people annually. Should she be concerned? Explain your reasoning.

   e. Assuming that the linear pattern continues, use the line given in part (d) to predict the size of the population in the United States in the next census.

2. In search of a topic for his science class project, Bill saw an interesting YouTube video in which dropping mint candies into bottles of a soda pop caused the soda pop to spurt immediately from the bottle. He wondered if the height of the spurt was linearly related to the number of mint candies that were used. He collected data using 1, 3, 5, and 10 mint candies. Then, he used two-liter bottles of a diet soda and measured the height of the spurt in centimeters. He tried each quantity of mint candies three times. His data are in the following table.

Number of Mint Candies	1	1	1	3	3	3	5	5	5	10	10	10
Height of Spurt (centimeters)	40	35	30	110	105	90	170	160	180	400	390	420

   a. Identify which variable is the independent variable and which is the dependent variable.

b. Draw a scatter plot that could be used to determine whether the relationship between height of spurt and number of mint candies appears to be linear.

c. Bill sees a slight curvature in the scatter plot, but he thinks that the relationship between the number of mint candies and the height of the spurt appears close enough to being linear, and he proceeds to draw a line. His eyeballed line goes through the mean of the three heights for three mint candies and the mean of the three heights for 10 candies. Bill calculates the equation of his eyeballed line to be

$$y = -27.617 + (43.095)x,$$

where the height of the spurt (y) in centimeters is based on the number of mint candies (x). Do you agree with this calculation? He rounded all of his calculations to three decimal places. Show your work.

d. In the context of this problem, interpret in words the slope and intercept for Bill's line. Does interpreting the intercept make sense in this context? Explain.

e. If the linear trend continues for greater numbers of mint candies, what do you predict the height of the spurt to be if 15 mint candies are used?

Lesson 12: Nonlinear Models in a Data Context

Classwork

Example 1: Growing Dahlias

A group of students wanted to determine whether or not compost is beneficial in plant growth. The students used the dahlia flower to study the effect of composting. They planted eight dahlias in a bed with no compost and another eight plants in a bed with compost. They measured the height of each plant over a 9-week period. They found the median growth height for each group of eight plants. The table below shows the results of the experiment for the dahlias grown in non-compost beds.

Week	Median Height in Non-Compost Bed (inches)
1	9.00
2	12.75
3	16.25
4	19.50
5	23.00
6	26.75
7	30.00
8	33.75
9	37.25

Exercises 1–15

1. On the grid below, construct a scatter plot of non-compost height versus week.

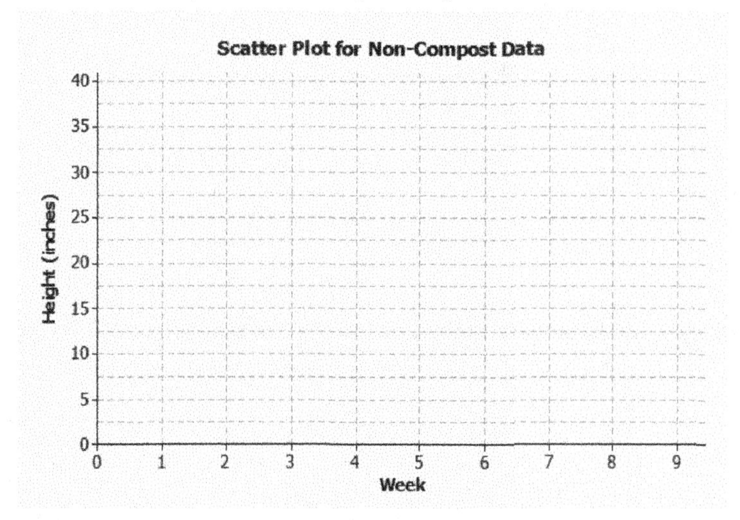

2. Draw a line that you think fits the data reasonably well.

3. Find the rate of change of your line. Interpret the rate of change in terms of growth (in height) over time.

4. Describe the growth (change in height) from week to week by subtracting the previous week's height from the current height. Record the weekly growth in the third column in the table below. The median growth for the dahlias from Week 1 to Week 2 was 3.75 inches (i.e., $12.75 - 9.00 = 3.75$).

Week	Median Height in Non-Compost Bed (inches)	Weekly Growth (inches)
1	9.00	—
2	12.75	3.75
3	16.25	
4	19.50	
5	23.00	
6	26.75	
7	30.00	
8	33.75	
9	37.25	

5. As the number of weeks increases, describe how the weekly growth is changing.

6. How does the growth each week compare to the slope of the line that you drew?

7. Estimate the median height of the dahlias at $8\frac{1}{2}$ weeks. Explain how you made your estimate.

The table below shows the results of the experiment for the dahlias grown in compost beds.

Week	Median Height in Compost Bed (inches)
1	10.00
2	13.50
3	17.75
4	21.50
5	30.50
6	40.50
7	65.00
8	80.50
9	91.50

8. Construct a scatter plot of height versus week on the grid below.

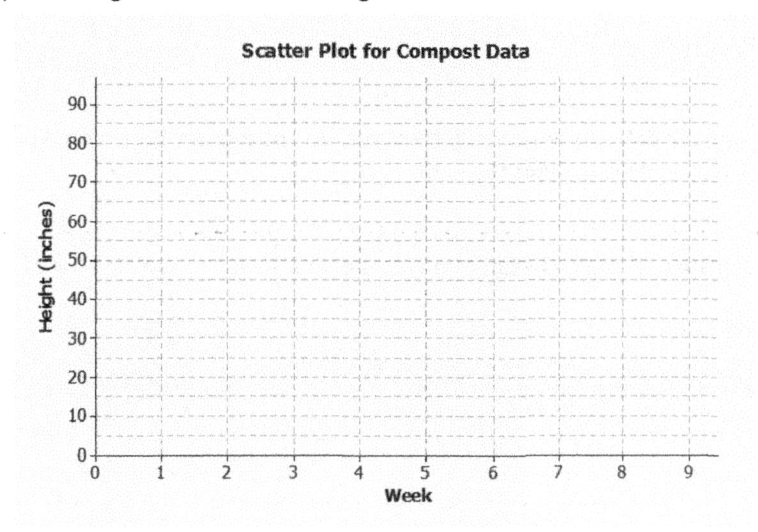

9. Do the data appear to form a linear pattern?

10. Describe the growth from week to week by subtracting the height from the previous week from the current height. Record the weekly growth in the third column in the table below. The median weekly growth for the dahlias from Week 1 to Week 2 is 3.5 inches. (i.e., $13.5 - 10 = 3.5$).

Week	Compost Height (inches)	Weekly Growth (inches)
1	10.00	–
2	13.50	3.50
3	17.75	
4	21.50	
5	30.50	
6	40.50	
7	65.00	
8	80.50	
9	91.50	

11. As the number of weeks increases, describe how the growth changes.

12. Sketch a curve through the data. When sketching a curve, do not connect the ordered pairs, but draw a smooth curve that you think reasonably describes the data.

13. Use the curve to estimate the median height of the dahlias at $8\frac{1}{2}$ weeks. Explain how you made your estimate.

14. How does the weekly growth of the dahlias in the compost beds compare to the weekly growth of the dahlias in the non-compost beds?

15. When there is a car accident, how do the investigators determine the speed of the cars involved? One way is to measure the skid marks left by the cars and use these lengths to estimate the speed.

The table below shows data collected from an experiment with a test car. The first column is the length of the skid mark (in feet), and the second column is the speed of the car (in miles per hour).

Skid-Mark Length (feet)	Speed (miles per hour)
5	10
17	20
65	40
105	50
205	70
265	80

Data Source: http://forensicdynamics.com/stopping-braking-distance-calculator

(Note: Data has been rounded.)

a. Construct a scatter plot of speed versus skid-mark length on the grid below.

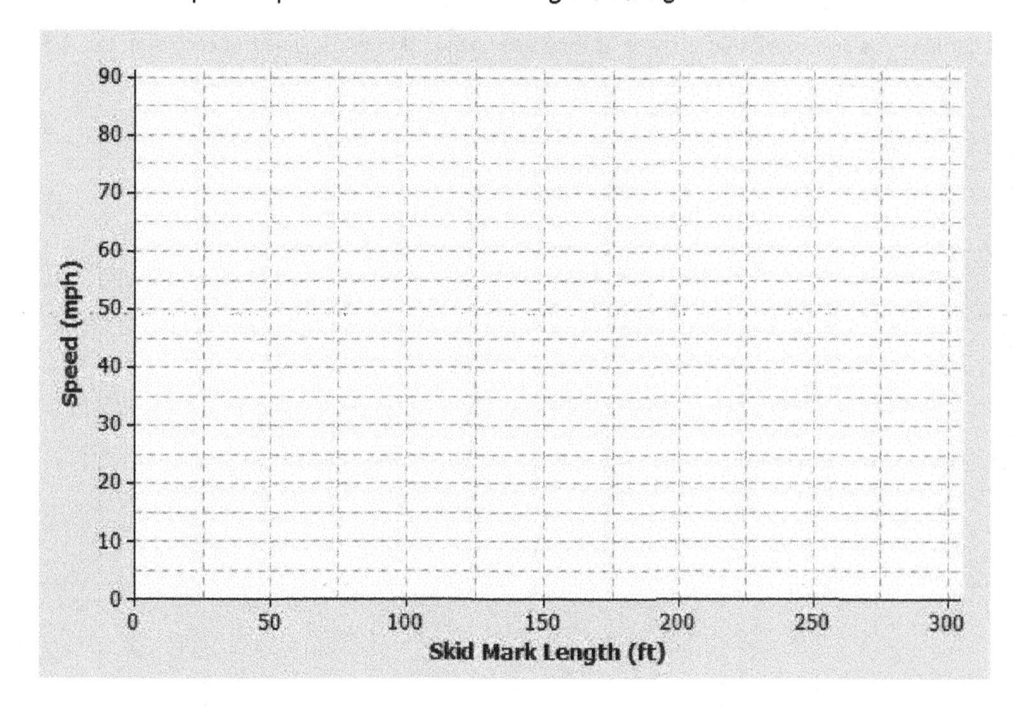

b. The relationship between speed and skid-mark length can be described by a curve. Sketch a curve through the data that best represents the relationship between skid-mark length and speed of the car. Remember to draw a smooth curve that does not just connect the ordered pairs.

c. If the car left a skid mark of 60 ft., what is an estimate for the speed of the car? Explain how you determined the estimate.

d. A car left a skid mark of 150 ft. Use the curve you sketched to estimate the speed at which the car was traveling.

e. If a car leaves a skid mark that is twice as long as another skid mark, was the car going twice as fast? Explain.

Lesson Summary

When data follow a linear pattern, they can be represented by a linear function whose rate of change can be used to answer questions about the data. When data do not follow a linear pattern, then there is no constant rate of change.

Problem Set

1. Once the brakes of the car have been applied, the car does not stop immediately. The distance that the car travels after the brakes have been applied is called the *braking distance*. The table below shows braking distance (how far the car travels once the brakes have been applied) and the speed of the car.

Speed (miles per hour)	Braking Distance (feet)
10	5
20	17
30	37
40	65
50	105
60	150
70	205
80	265

Data Source: http://forensicdynamics.com/stopping-braking-distance-calculator

(Note: Data has been rounded.)

a. Construct a scatter plot of braking distance versus speed on the grid below.

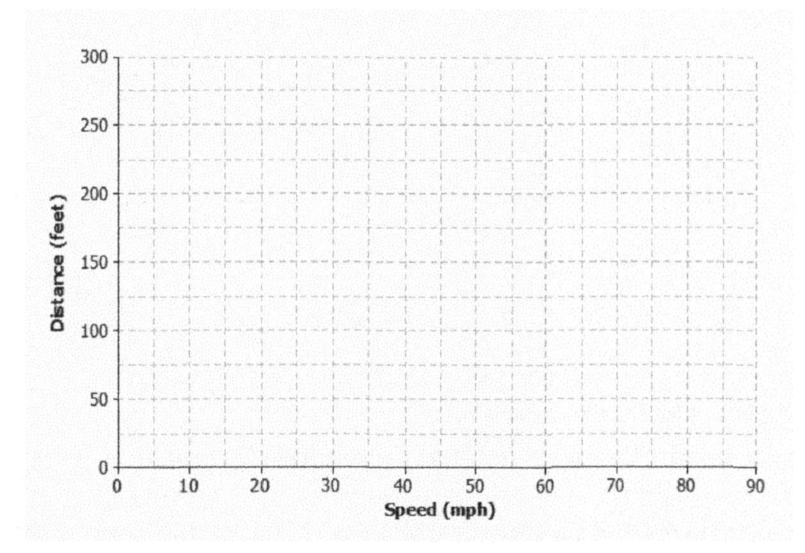

b. Find the amount of additional distance a car would travel after braking for each speed increase of 10 mph. Record your answers in the table below.

Speed (miles per hour)	Braking Distance (feet)	Amount of Distance Increase
10	5	—
20	17	
30	37	
40	65	
50	105	
60	150	
70	205	
80	265	

c. Based on the table, do you think the data follow a linear pattern? Explain your answer.

d. Describe how the distance it takes a car to stop changes as the speed of the car increases.

e. Sketch a smooth curve that you think describes the relationship between braking distance and speed.

f. Estimate braking distance for a car traveling at 52 mph. Estimate braking distance for a car traveling at 75 mph. Explain how you made your estimates.

2. The scatter plot below shows the relationship between cost (in dollars) and radius length (in meters) of fertilizing different-sized circular fields. The curve shown was drawn to describe the relationship between cost and radius.

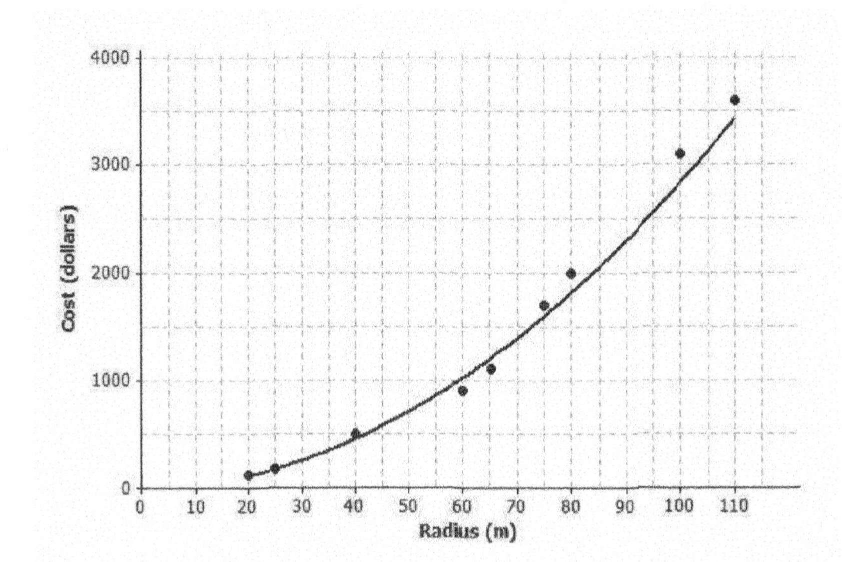

a. Is the curve a good fit for the data? Explain.

b. Use the curve to estimate the cost for fertilizing a circular field of radius 30 m. Explain how you made your estimate.

c. Estimate the radius of the field if the fertilizing cost was $\$2,500$. Explain how you made your estimate.

3. Suppose a dolphin is fitted with a GPS that monitors its position in relationship to a research ship. The table below contains the time (in seconds) after the dolphin is released from the ship and the distance (in feet) the dolphin is from the research ship.

Time (seconds)	Distance from the Ship (feet)	Increase in Distance from the Ship
0	0	–
50	85	
100	190	
150	398	
200	577	
250	853	
300	1,122	

a. Construct a scatter plot of distance versus time on the grid below.

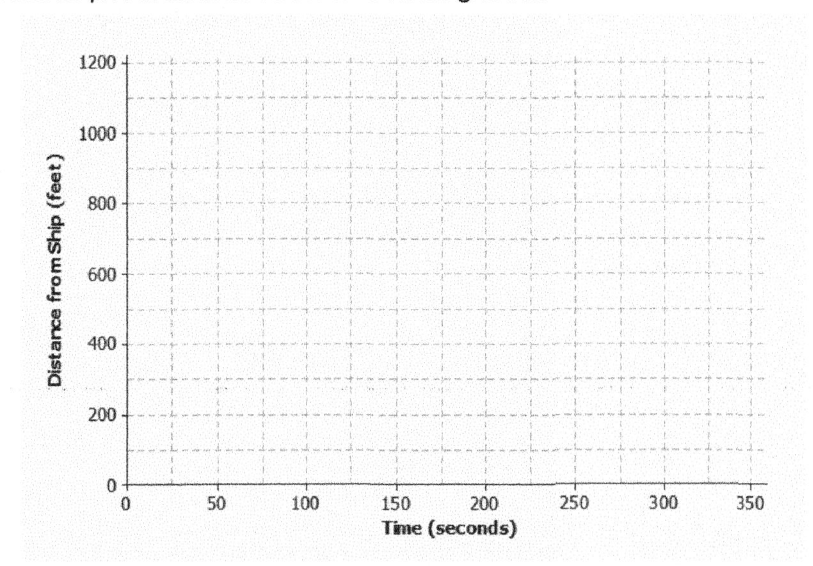

b. Find the additional distance the dolphin traveled for each increase of 50 seconds. Record your answers in the table above.

c. Based on the table, do you think that the data follow a linear pattern? Explain your answer.

d. Describe how the distance that the dolphin is from the ship changes as the time increases.

e. Sketch a smooth curve that you think fits the data reasonably well.

f. Estimate how far the dolphin will be from the ship after 180 seconds. Explain how you made your estimate.

This page intentionally left blank

Lesson 13: Summarizing Bivariate Categorical Data in a Two-Way Table

Classwork

Exercises 1–8

On an upcoming field day at school, the principal wants to provide ice cream during lunch. She offers three flavors: chocolate, strawberry, and vanilla. She selected your class to complete a survey to help her determine how much of each flavor to buy.

1. Answer the following question. Wait for your teacher to count how many students selected each flavor. Then, record the class totals for each flavor in the table below.

   "Which of the following three ice cream flavors is your favorite: chocolate, strawberry, or vanilla?"

Ice Cream Flavor	Chocolate	Strawberry	Vanilla	Total
Number of Students				

2. Which ice cream flavor do most students prefer?

3. Which ice cream flavor do the fewest students prefer?

4. What percentage of students preferred each flavor? Round to the nearest tenth of a percent.

5. Do the numbers in the table in Exercise 1 summarize data on a categorical variable or a numerical variable?

6. Do the students in your class represent a random sample of all the students in your school? Why or why not? Discuss this with your neighbor.

7. Is your class representative of all the other classes at your school? Why or why not? Discuss this with your neighbor.

8. Do you think the principal will get an accurate estimate of the proportion of students who prefer each ice cream flavor for the whole school using only your class? Why or why not? Discuss this with your neighbor.

Example 1

Students in a different class were asked the same question about their favorite ice cream flavors. The table below shows the ice cream flavors and the number of students who chose each flavor for that particular class. This table is called a *one-way frequency table* because it shows the counts of a univariate categorical variable.

This is the univariate categorical variable. ⟶

These are the counts for each category. ⟶

Ice Cream Flavor	Chocolate	Strawberry	Vanilla	Total
Number of Students	11	4	10	25

We compute the relative frequency for each ice cream flavor by dividing the count by the total number of observations.

$$\text{relative frequency} = \frac{\text{count for a category}}{\text{total number of observations}}$$

Since 11 out of 25 students answered *chocolate*, the relative frequency would be $\frac{11}{25} \approx 0.44$. This relative frequency shows that 44% of the class prefers chocolate ice cream. In other words, the relative frequency is the proportional value that each category is of the whole.

Exercises 9–10

Use the table for the preferred ice cream flavors from the class in Example 1 to answer the following questions.

9. What is the relative frequency for the category *strawberry*?

10. Write a sentence interpreting the relative frequency value in the context of strawberry ice cream preference.

Example 2

The principal also wondered if boys and girls have different favorite ice cream flavors. She decided to redo the survey by taking a random sample of students from the school and recording both their favorite ice cream flavors and their genders. She asked the following two questions:

- "Which of the following ice cream flavors is your favorite: chocolate, strawberry, or vanilla?"
- "What is your gender: male or female?"

The results of the survey are as follows:

- Of the 30 students who prefer chocolate ice cream, 22 are males.
- Of the 25 students who prefer strawberry ice cream, 15 are females.
- Of the 27 students who prefer vanilla ice cream, 13 are males.

The values of two variables, which were ice cream flavor and gender, were recorded in this survey. Since both of the variables are categorical, the data are bivariate categorical data.

Exercises 11–17

11. Can we display these data in a one-way frequency table? Why or why not?

12. Summarize the results of the second survey of favorite ice cream flavors in the following table:

		Favorite Ice Cream Flavor			
		Chocolate	Strawberry	Vanilla	Total
Gender	Male				
	Female				
	Total				

13. Calculate the relative frequencies of the data in the table in Exercise 12, and write them in the following table.

		Favorite Ice Cream Flavor			
		Chocolate	Strawberry	Vanilla	Total
Gender	Male				
	Female				
	Total				

Use the relative frequency values in the table to answer the following questions:

14. What is the proportion of the students who prefer chocolate ice cream?

15. What is the proportion of students who are female and prefer vanilla ice cream?

16. Write a sentence explaining the meaning of the approximate relative frequency 0.55.

17. Write a sentence explaining the meaning of the approximate relative frequency 0.10.

Example 3

In the previous exercises, you used the total number of students to calculate relative frequencies. These relative frequencies were the proportion of the whole group who answered the survey a certain way. Sometimes we use row or column totals to calculate relative frequencies. We call these *row relative frequencies* or *column relative frequencies*.

Below is the two-way frequency table for your reference. To calculate "the proportion of male students who prefer chocolate ice cream," divide the 22 male students who preferred chocolate ice cream by the total of 45 male students. This proportion is $\frac{22}{45} = 0.49$. Notice that you used the row total to make this calculation. This is a row relative frequency.

		Favorite Ice Cream Flavor			
		Chocolate	**Strawberry**	**Vanilla**	**Total**
Gender	**Male**	22	10	13	45
	Female	8	15	14	37
	Total	30	25	27	82

Exercises 18–22

In Exercise 13, you used the total number of students to calculate relative frequencies. These relative frequencies were the proportion of the whole group who answered the survey a certain way.

18. Suppose you are interested in the proportion of male students who prefer chocolate ice cream. How is this value different from "the proportion of students who are male and prefer chocolate ice cream"? Discuss this with your neighbor.

19. Use the table provided in Example 3 to calculate the following relative frequencies.
    a. What proportion of students who prefer vanilla ice cream are female?

b. What proportion of male students prefer strawberry ice cream? Write a sentence explaining the meaning of this proportion in context of this problem.

c. What proportion of female students prefer strawberry ice cream?

d. What proportion of students who prefer strawberry ice cream are female?

20. A student is selected at random from this school. What would you predict this student's favorite ice cream to be? Explain why you chose this flavor.

21. Suppose the randomly selected student is male. What would you predict his favorite flavor of ice cream to be? Explain why you chose this flavor.

22. Suppose the randomly selected student is female. What would you predict her favorite flavor of ice cream to be? Explain why you chose this flavor.

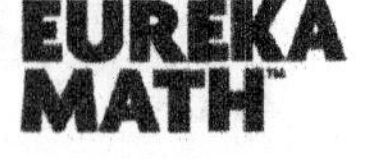

Lesson Summary

- Univariate categorical data are displayed in a one-way frequency table.
- Bivariate categorical data are displayed in a two-way frequency table.
- *Relative frequency* is the frequency divided by a total $\left(\frac{\text{frequency}}{\text{total}}\right)$.
- A *cell relative frequency* is a cell frequency divided by the total number of observations.
- A *row relative frequency* is a cell frequency divided by the row total.
- A *column relative frequency* is a cell frequency divided by the column total.

Problem Set

Every student at Abigail Douglas Middle School is enrolled in exactly one extracurricular activity. The school counselor recorded data on extracurricular activity and gender for all 254 eighth-grade students at the school.

The counselor's findings for the 254 eighth-grade students are the following:

- Of the 80 students enrolled in band, 42 are male.Of the 65 students enrolled in choir, 20 are male.
- Of the 88 students enrolled in sports, 30 are female.
- Of the 21 students enrolled in art, 9 are female.

1. Complete the table below.

		Extracurricular Activities				
		Band	Choir	Sports	Art	Total
Gender	Female					
	Male					
	Total					

2. Write a sentence explaining the meaning of the frequency 38 in this table.

Use the table provided above to calculate the following relative frequencies.

3. What proportion of students are male and enrolled in choir?

4. What proportion of students are enrolled in a musical extracurricular activity (i.e., band or choir)?

5. What proportion of male students are enrolled in sports?

6. What proportion of students enrolled in sports are male?

Pregnant women often undergo ultrasound tests to monitor their babies' health. These tests can also be used to predict the gender of the babies, but these predictions are not always accurate. Data on the gender predicted by ultrasound and the actual gender of the baby for 1,000 babies are summarized in the two-way table below.

		Predicted Gender	
		Female	Male
Actual Gender	Female	432	48
	Male	130	390

7. Write a sentence explaining the meaning of the frequency 130 in this table.

Use the table provided above to calculate the following relative frequencies.

8. What is the proportion of babies who were predicted to be male but were actually female?

9. What is the proportion of incorrect ultrasound gender predictions?

10. For babies predicted to be female, what proportion of the predictions were correct?

11. For babies predicted to be male, what proportion of the predictions were correct?

Lesson 14: Association Between Categorical Variables

Classwork

Example 1

Suppose a random group of people are surveyed about their use of smartphones. The results of the survey are summarized in the tables below.

Smartphone Use and Gender

	Use a Smartphone	Do Not Use a Smartphone	Total
Male	30	10	40
Female	45	15	60
Total	75	25	100

Smartphone Use and Age

	Use a Smartphone	Do Not Use a Smartphone	Total
Under 40 Years of Age	45	5	50
40 Years of Age or Older	30	20	50
Total	75	25	100

Example 2

Suppose a sample of 400 participants (teachers and students) was randomly selected from the middle schools and high schools in a large city. These participants responded to the following question:

Which type of movie do you prefer to watch?

1. Action (*The Avengers, Man of Steel*, etc.)
2. Drama (*42 (The Jackie Robinson Story), The Great Gatsby*, etc.)
3. Science Fiction (*Star Trek into Darkness, World War Z*, etc.)
4. Comedy (*Monsters University, Despicable Me 2*, etc.)

Movie preference and status (teacher or student) were recorded for each participant.

Exercises 1–7

1. Two variables were recorded. Are these variables categorical or numerical?

2. The results of the survey are summarized in the table below.

	Movie Preference				
	Action	**Drama**	**Science Fiction**	**Comedy**	**Total**
Student	120	60	30	90	300
Teacher	40	20	10	30	100
Total	160	80	40	120	400

a. What proportion of participants who are teachers prefer action movies?

b. What proportion of participants who are teachers prefer drama movies?

c. What proportion of participants who are teachers prefer science fiction movies?

d. What proportion of participants who are teachers prefer comedy movies?

The answers to Exercise 2 are called *row relative frequencies*. Notice that you divided each cell frequency in the Teacher row by the total for that row. Below is a blank relative frequency table.

Table of Row Relative Frequencies

	Movie Preference			
	Action	**Drama**	**Science Fiction**	**Comedy**
Student				
Teacher	(a)	(b)	(c)	(d)

Write your answers from Exercise 2 in the indicated cells in the table above.

3. Find the row relative frequencies for the Student row. Write your answers in the table above.
   a. What proportion of participants who are students prefer action movies?
   b. What proportion of participants who are students prefer drama movies?
   c. What proportion of participants who are students prefer science fiction movies?
   d. What proportion of participants who are students prefer comedy movies?

4. Is a participant's status (i.e., teacher or student) related to what type of movie he would prefer to watch? Why or why not? Discuss this with your group.

5. What does it mean when we say that there is *no association* between two variables? Discuss this with your group.

6. Notice that the row relative frequencies for each movie type are the same for both the Teacher and Student rows. When this happens, we say that the two variables, movie preference and status (student or teacher), are *not* associated. Another way of thinking about this is to say that knowing if a participant is a teacher (or a student) provides no information about his movie preference.

   What does it mean if row relative frequencies are not the same for all rows of a two-way table?

7. You can also evaluate whether two variables are associated by looking at column relative frequencies instead of row relative frequencies. A column relative frequency is a cell frequency divided by the corresponding column total. For example, the column relative frequency for the Student/Action cell is $\frac{120}{160} = 0.75$.

   a. Calculate the other column relative frequencies, and write them in the table below.

Table of Column Relative Frequencies

	Movie Preference			
	Action	**Drama**	**Science Fiction**	**Comedy**
Student				
Teacher				

   b. What do you notice about the column relative frequencies for the four columns?

   c. What would you conclude about association based on the column relative frequencies?

Example 3

In the survey described in Example 2, gender for each of the 400 participants was also recorded. Some results of the survey are given below:

- 160 participants preferred action movies.
- 80 participants preferred drama movies.
- 40 participants preferred science fiction movies.
- 240 participants were females.
- 78 female participants preferred drama movies.
- 32 male participants preferred science fiction movies.
- 60 female participants preferred action movies.

Exercises 8–15

Use the results from Example 3 to answer the following questions. Be sure to discuss these questions with your group members.

8. Complete the two-way frequency table that summarizes the data on movie preference and gender.

	Movie Preference				
	Action	Drama	Science Fiction	Comedy	Total
Female					
Male					
Total					

9. What proportion of the participants are female?

10. If there was no association between gender and movie preference, should you expect more females than males or fewer females than males to prefer action movies? Explain.

11. Make a table of row relative frequencies of each movie type for the Male row and the Female row. Refer to Exercises 2–4 to review how to complete the table below.

	Movie Preference			
	Action	Drama	Science Fiction	Comedy
Female				
Male				

Suppose that you randomly pick 1 of the 400 participants. Use the table of row relative frequencies on the previous page to answer the following questions.

12. If you had to predict what type of movie this person chose, what would you predict? Explain why you made this choice.

13. If you know that the randomly selected participant is female, would you predict that her favorite type of movie is action? If not, what would you predict, and why?

14. If knowing the value of one of the variables provides information about the value of the other variable, then there is an association between the two variables.

    Is there an association between the variables gender and movie preference? Explain.

15. What can be said when two variables are associated? Read the following sentences. Decide if each sentence is a correct statement based upon the survey data. If it is not correct, explain why not.

    a. More females than males participated in the survey.

    b. Males tend to prefer action and science fiction movies.

    c. Being female causes one to prefer drama movies.

Lesson Summary

- Saying that two variables *are not* associated means that knowing the value of one variable provides no information about the value of the other variable.
- Saying that two variables *are* associated means that knowing the value of one variable provides information about the value of the other variable.
- To determine if two variables are associated, calculate row relative frequencies. If the row relative frequencies are about the same for all of the rows, it is reasonable to say that there is no association between the two variables that define the table.
- Another way to decide if there is an association between two categorical variables is to calculate column relative frequencies. If the column relative frequencies are about the same for all of the columns, it is reasonable to say that there is no association between the two variables that define the table.
- If the row relative frequencies are quite different for some of the rows, it is reasonable to say that there is an association between the two variables that define the table.

Problem Set

A sample of 200 middle school students was randomly selected from the middle schools in a large city. Answers to several survey questions were recorded for each student. The tables below summarize the results of the survey.

For each table, calculate the row relative frequencies for the Female row and for the Male row. Write the row relative frequencies beside the corresponding frequencies in each table below.

1. This table summarizes the results of the survey data for the two variables, gender and which sport the students prefer to play. Is there an association between gender and which sport the students prefer to play? Explain.

		Sport				
		Football	**Basketball**	**Volleyball**	**Soccer**	**Total**
Gender	**Female**	2	29	28	38	97
	Male	35	26	8	24	103
	Total	37	65	36	62	200

2. This table summarizes the results of the survey data for the two variables, gender and the students' T-shirt sizes. Is there an association between gender and T-shirt size? Explain.

		School T-Shirt Sizes				
		Small	Medium	Large	X-Large	Total
Gender	Female	47	35	13	2	97
	Male	11	41	42	9	103
	Total	58	76	55	11	200

3. This table summarizes the results of the survey data for the two variables, gender and favorite type of music. Is there an association between gender and favorite type of music? Explain

		Favorite Type of Music				
		Pop	Hip-Hop	Alternative	Country	Total
Gender	Female	35	28	11	23	97
	Male	37	30	13	23	103
	Total	72	58	24	46	200

Student Edition

Eureka Math
Grade 8
Module 7

Special thanks go to the Gordan A. Cain Center and to the Department of Mathematics at Louisiana State University for their support in the development of *Eureka Math*.

Published by the non-profit Great Minds

Printed in the U.S.A.
This book may be purchased from the publisher at eureka-math.org
10 9 8 7 6 5 4 3 2 1

ISBN 978-1-63255-322-5

Lesson 1: The Pythagorean Theorem

Classwork

Note: The figures in this lesson are not drawn to scale.

Example 1

Write an equation that allows you to determine the length of the unknown side of the right triangle.

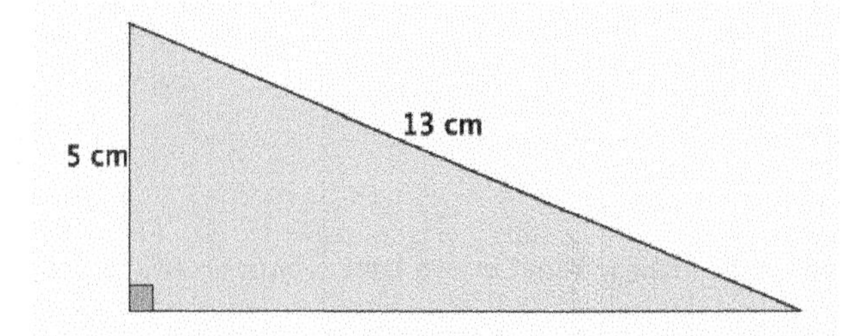

Example 2

Write an equation that allows you to determine the length of the unknown side of the right triangle.

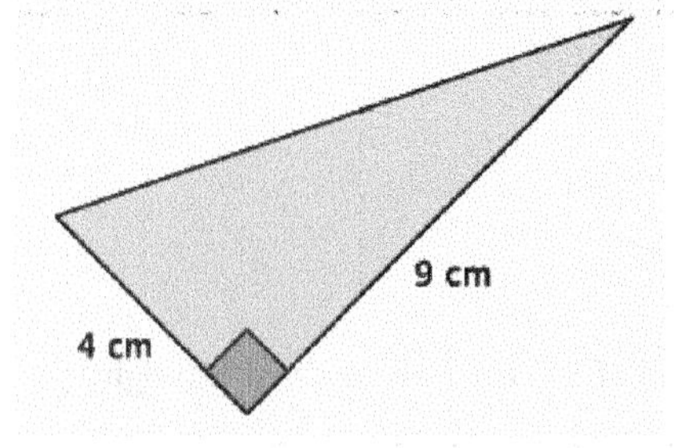

Example 3

Write an equation to determine the length of the unknown side of the right triangle.

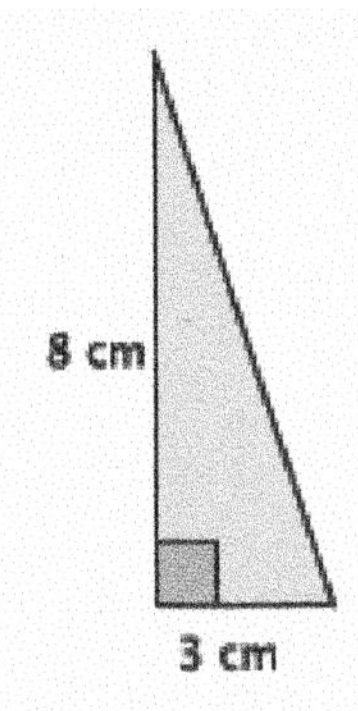

Example 4

In the figure below, we have an equilateral triangle with a height of 10 inches. What do we know about an equilateral triangle?

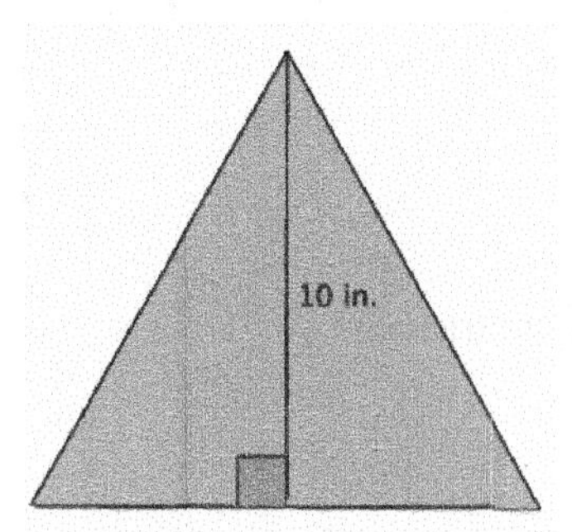

Exercises

1. Use the Pythagorean theorem to find a whole number estimate of the length of the unknown side of the right triangle. Explain why your estimate makes sense.

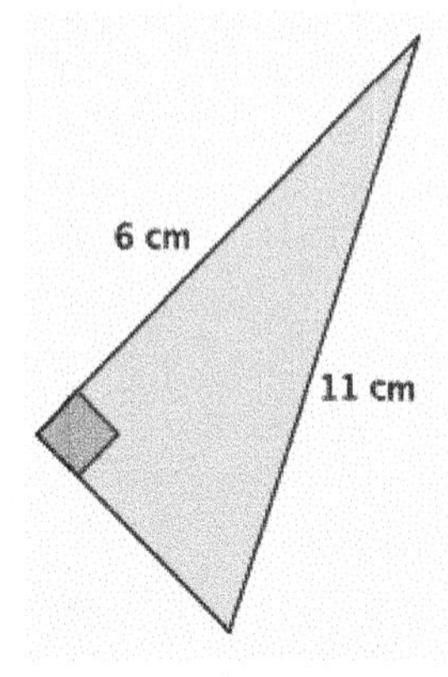

2. Use the Pythagorean theorem to find a whole number estimate of the length of the unknown side of the right triangle. Explain why your estimate makes sense.

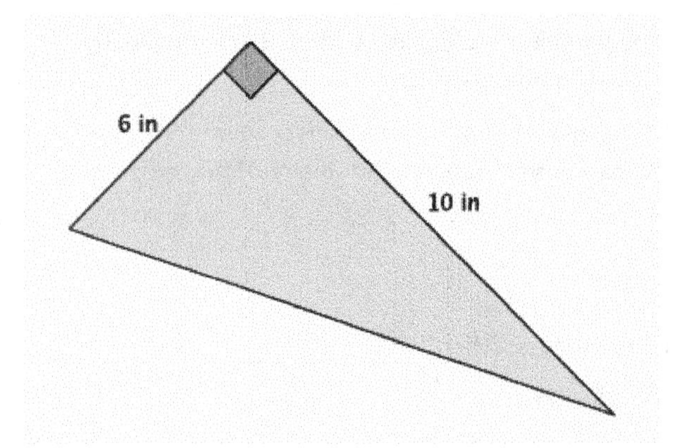

3. Use the Pythagorean theorem to find a whole number estimate of the length of the unknown side of the right triangle. Explain why your estimate makes sense.

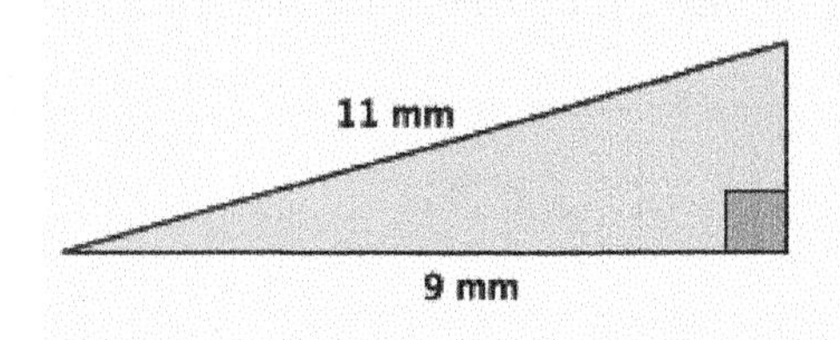

Lesson Summary

Perfect square numbers are those that are a product of an integer factor multiplied by itself. For example, the number 25 is a perfect square number because it is the product of 5 multiplied by 5.

When the square of the length of an unknown side of a right triangle is not equal to a perfect square, you can estimate the length as a whole number by determining which two perfect squares the square of the length is between.

Example:

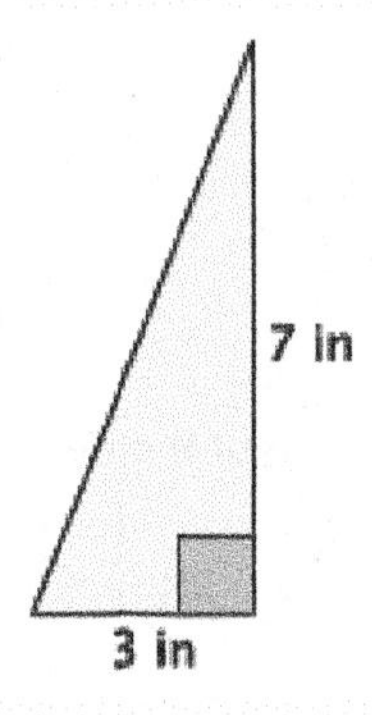

Let c in. represent the length of the hypotenuse. Then,

$$3^2 + 7^2 = c^2$$
$$9 + 49 = c^2$$
$$58 = c^2.$$

The number 58 is not a perfect square, but it is between the perfect squares 49 and 64. Therefore, the length of the hypotenuse is between 7 in. and 8 in. but closer to 8 in. because 58 is closer to the perfect square 64 than it is to the perfect square 49.

Problem Set

1. Use the Pythagorean theorem to estimate the length of the unknown side of the right triangle. Explain why your estimate makes sense.

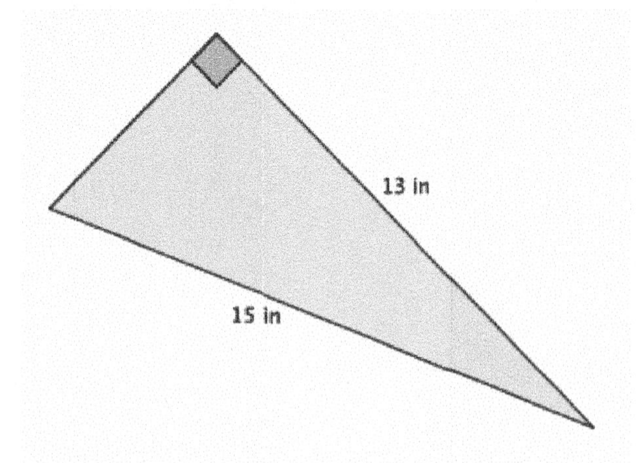

2. Use the Pythagorean theorem to estimate the length of the unknown side of the right triangle. Explain why your estimate makes sense.

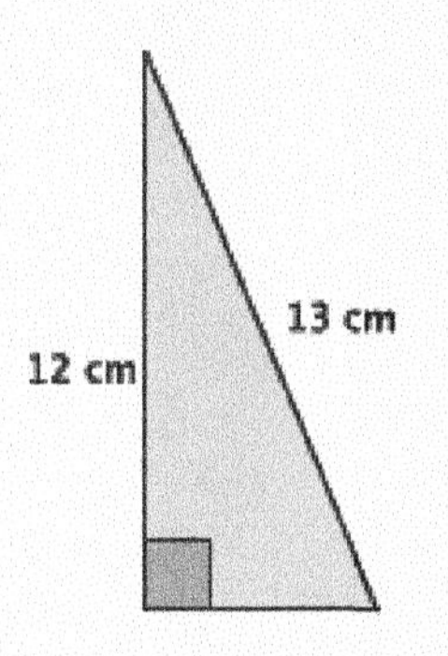

3. Use the Pythagorean theorem to estimate the length of the unknown side of the right triangle. Explain why your estimate makes sense.

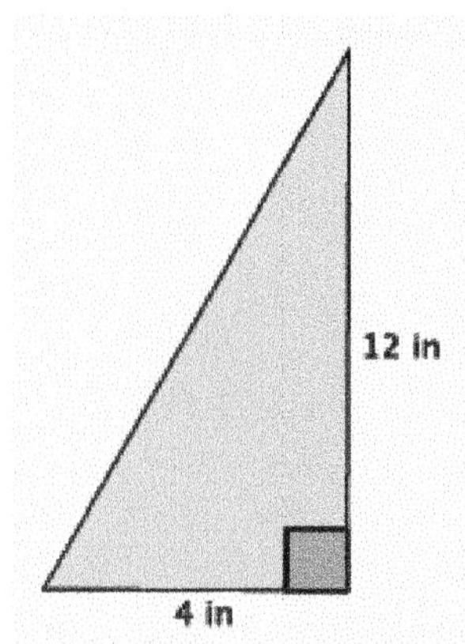

4. Use the Pythagorean theorem to estimate the length of the unknown side of the right triangle. Explain why your estimate makes sense.

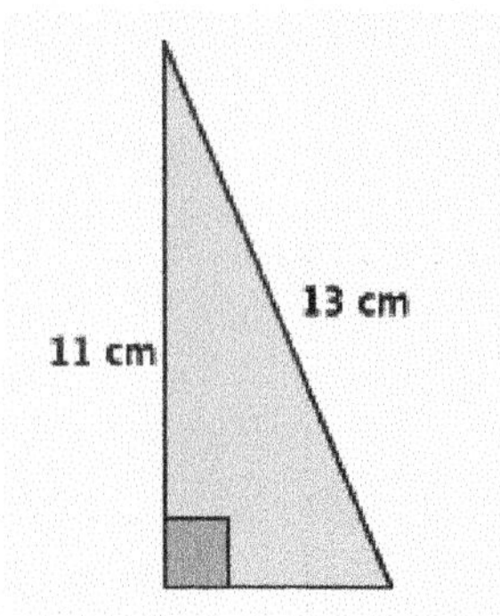

5. Use the Pythagorean theorem to estimate the length of the unknown side of the right triangle. Explain why your estimate makes sense.

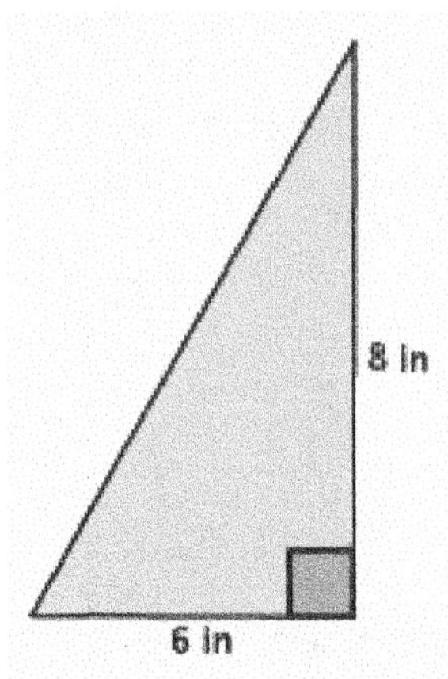

6. Determine the length of the unknown side of the right triangle. Explain how you know your answer is correct.

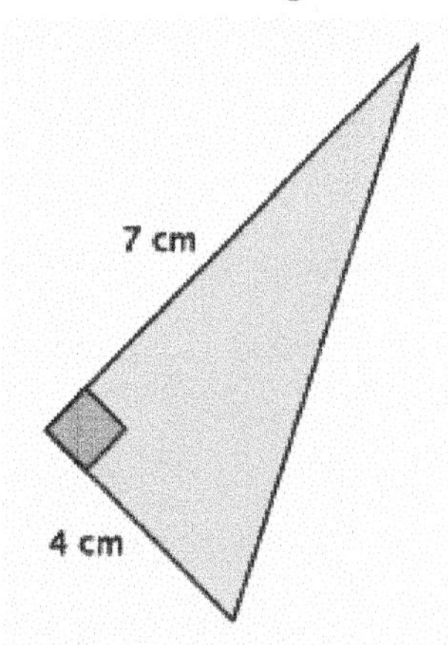

7. Use the Pythagorean theorem to estimate the length of the unknown side of the right triangle. Explain why your estimate makes sense.

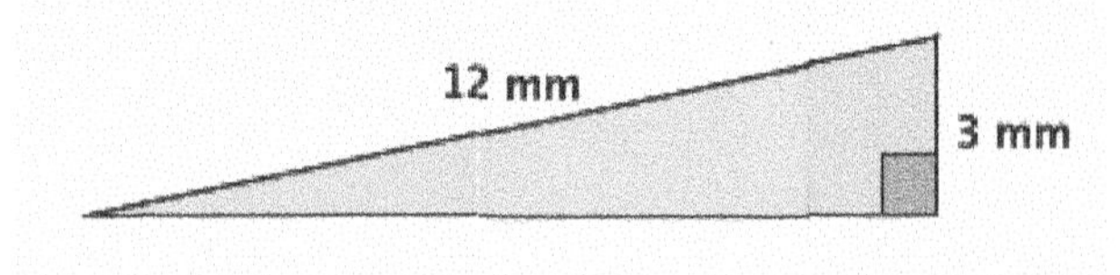

8. The triangle below is an isosceles triangle. Use what you know about the Pythagorean theorem to determine the approximate length of the base of the isosceles triangle.

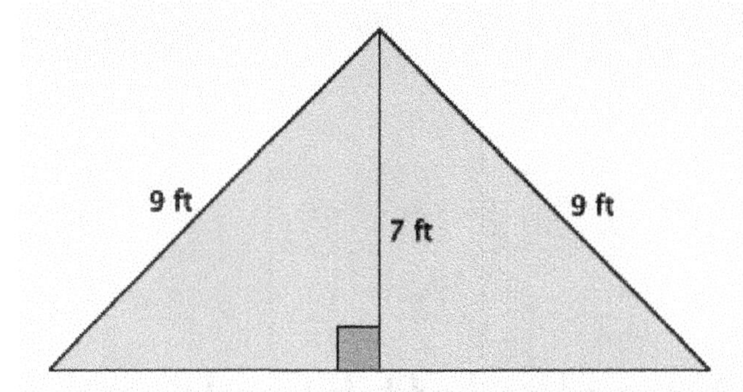

9. Give an estimate for the area of the triangle shown below. Explain why it is a good estimate.

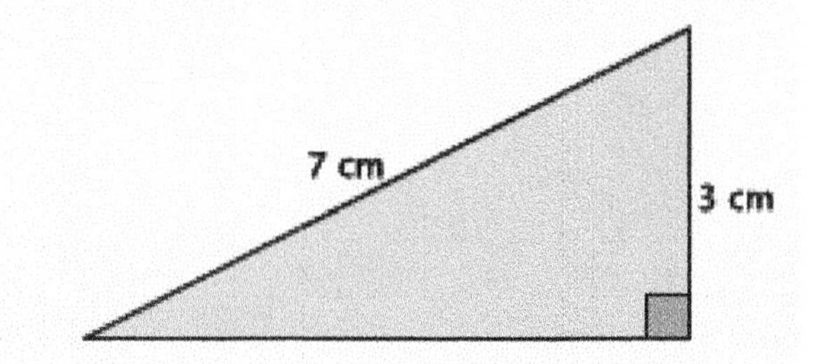

This page intentionally left blank

Lesson 2: Square Roots

Classwork

Exercises 1–4

1. Determine the positive square root of 81, if it exists. Explain.

2. Determine the positive square root of 225, if it exists. Explain.

3. Determine the positive square root of -36, if it exists. Explain.

4. Determine the positive square root of 49, if it exists. Explain.

Discussion

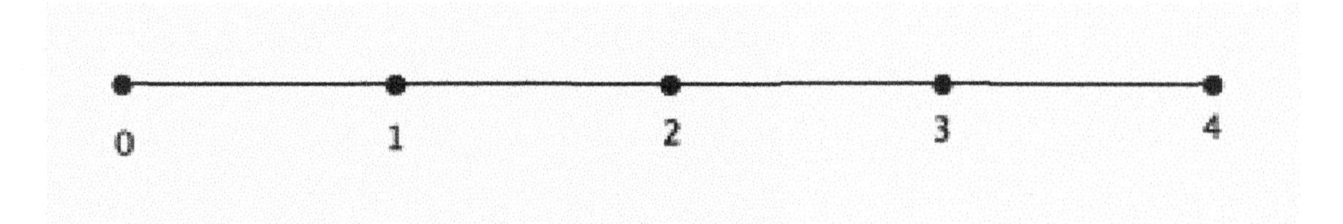

Exercises 5–9

Determine the positive square root of the number given. If the number is not a perfect square, determine which whole number the square root would be closest to, and then use *guess and check* to give an approximate answer to one or two decimal places.

5. $\sqrt{49}$

6. $\sqrt{62}$

7. $\sqrt{122}$

8. $\sqrt{400}$

9. Which of the numbers in Exercises 5–8 are not perfect squares? Explain.

Lesson Summary

A positive number whose square is equal to a positive number b is denoted by the symbol $\sqrt{b}$. The symbol $\sqrt{b}$ automatically denotes a positive number. For example, $\sqrt{4}$ is always 2, not -2. The number $\sqrt{b}$ is called a *positive square root of* b.

The square root of a perfect square of a whole number is that whole number. However, there are many whole numbers that are not perfect squares.

Problem Set

Determine the positive square root of the number given. If the number is not a perfect square, determine the integer to which the square root would be closest.

1. $\sqrt{169}$

2. $\sqrt{256}$

3. $\sqrt{81}$

4. $\sqrt{147}$

5. $\sqrt{8}$

6. Which of the numbers in Problems 1–5 are not perfect squares? Explain.

7. Place the following list of numbers in their approximate locations on a number line.

$$\sqrt{32}, \sqrt{12}, \sqrt{27}, \sqrt{18}, \sqrt{23}, \text{ and } \sqrt{50}$$

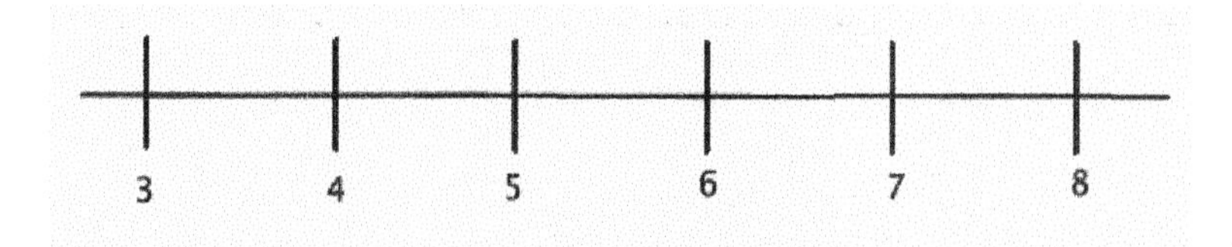

8. Between which two integers will $\sqrt{45}$ be located? Explain how you know.

This page intentionally left blank

Lesson 3: Existence and Uniqueness of Square Roots and Cube Roots

Classwork

Opening

The numbers in each column are related. Your goal is to determine how they are related, determine which numbers belong in the blank parts of the columns, and write an explanation for how you know the numbers belong there.

Find the Rule Part 1

1	1
2	
3	9
	81
11	121
15	
	49
10	
12	
	169
m	
	n

Find the Rule Part 2

1	1
2	
3	27
	125
6	216
11	
	64
10	
7	
	2,744
p	
	q

Exercises

Find the positive value of x that makes each equation true. Check your solution.

1. $x^2 = 169$

   a. Explain the first step in solving this equation.

   b. Solve the equation, and check your answer.

2. A square-shaped park has an area of 324 yd^2. What are the dimensions of the park? Write and solve an equation.

3. $625 = x^2$

4. A cube has a volume of 27 in^3. What is the measure of one of its sides? Write and solve an equation.

5. What positive value of x makes the following equation true: $x^2 = 64$? Explain.

6. What positive value of x makes the following equation true: $x^3 = 64$? Explain.

7. Find the positive value of x that makes the equation true: $x^2 = 256^{-1}$.

8. Find the positive value of x that makes the equation true: $x^3 = 343^{-1}$.

9. Is 6 a solution to the equation $x^2 - 4 = 5x$? Explain why or why not.

Lesson Summary

The symbol $\sqrt[n]{\ }$ is called a *radical*. An equation that contains that symbol is referred to as a *radical equation*. So far, we have only worked with square roots (i.e., $n = 2$). Technically, we would denote a positive square root as $\sqrt[2]{\ }$, but it is understood that the symbol $\sqrt{\ }$ alone represents a positive square root.

When $n = 3$, then the symbol $\sqrt[3]{\ }$ is used to denote the cube root of a number. Since $x^3 = x \cdot x \cdot x$, the cube root of x^3 is x (i.e., $\sqrt[3]{x^3} = x$).

The square or cube root of a positive number exists, and there can be only one positive square root or one cube root of the number.

Problem Set

Find the positive value of x that makes each equation true. Check your solution.

1. What positive value of x makes the following equation true: $x^2 = 289$? Explain.

2. A square-shaped park has an area of 400 yd^2. What are the dimensions of the park? Write and solve an equation.

3. A cube has a volume of 64 in^3. What is the measure of one of its sides? Write and solve an equation.

4. What positive value of x makes the following equation true: $125 = x^3$? Explain.

5. Find the positive value of x that makes the equation true: $x^2 = 441^{-1}$.
   a. Explain the first step in solving this equation.
   b. Solve and check your solution.

6. Find the positive value of x that makes the equation true: $x^3 = 125^{-1}$.

7. The area of a square is 196 in^2. What is the length of one side of the square? Write and solve an equation, and then check your solution.

8. The volume of a cube is 729 cm^3. What is the length of one side of the cube? Write and solve an equation, and then check your solution.

9. What positive value of x would make the following equation true: $19 + x^2 = 68$?

Lesson 4: Simplifying Square Roots

Classwork

Opening Exercise

a.

i. What does $\sqrt{16}$ equal?

ii. What does 4×4 equal?

iii. Does $\sqrt{16} = \sqrt{4 \times 4}$?

b.

i. What does $\sqrt{36}$ equal?

ii. What does 6×6 equal?

iii. Does $\sqrt{36} = \sqrt{6 \times 6}$?

c.

i. What does $\sqrt{121}$ equal?

ii. What does 11×11 equal?

iii. Does $\sqrt{121} = \sqrt{11 \times 11}$?

d.

i. What does $\sqrt{81}$ equal?

ii. What does 9×9 equal?

iii. Does $\sqrt{81} = \sqrt{9 \times 9}$?

e. Rewrite $\sqrt{20}$ using at least one perfect square factor.

f. Rewrite $\sqrt{28}$ using at least one perfect square factor.

Example 1

Simplify the square root as much as possible.

$\sqrt{50} =$

Example 2

Simplify the square root as much as possible.

$\sqrt{28} =$

Exercises 1–4

Simplify the square roots as much as possible.

1. $\sqrt{18}$

2. $\sqrt{44}$

3. $\sqrt{169}$

4. $\sqrt{75}$

Example 3

Simplify the square root as much as possible.

$\sqrt{128} =$

Example 4

Simplify the square root as much as possible.

$\sqrt{288} =$

Exercises 5–8

5. Simplify $\sqrt{108}$.

6. Simplify $\sqrt{250}$.

7. Simplify $\sqrt{200}$.

8. Simplify $\sqrt{504}$.

Lesson Summary

Square roots of some non-perfect squares can be simplified by using the factors of the number. Any perfect square factors of a number can be simplified.

For example:

$$\begin{aligned}\sqrt{72} &= \sqrt{36 \times 2} \\ &= \sqrt{36} \times \sqrt{2} \\ &= \sqrt{6^2} \times \sqrt{2} \\ &= 6 \times \sqrt{2} \\ &= 6\sqrt{2}\end{aligned}$$

Problem Set

Simplify each of the square roots in Problems 1–5 as much as possible.

1. $\sqrt{98}$

2. $\sqrt{54}$

3. $\sqrt{144}$

4. $\sqrt{512}$

5. $\sqrt{756}$

6. What is the length of the unknown side of the right triangle? Simplify your answer, if possible.

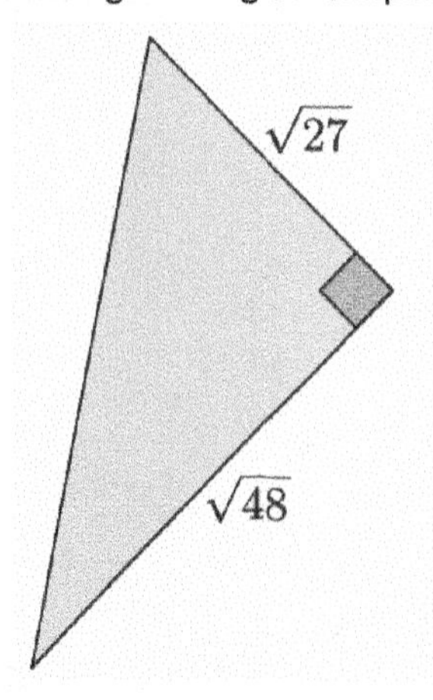

7. What is the length of the unknown side of the right triangle? Simplify your answer, if possible.

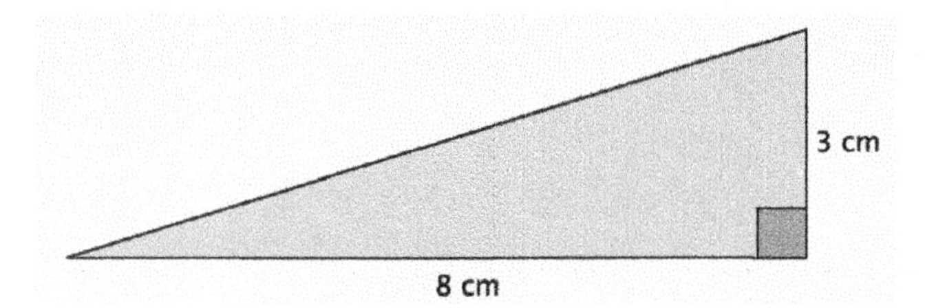

8. What is the length of the unknown side of the right triangle? Simplify your answer, if possible.

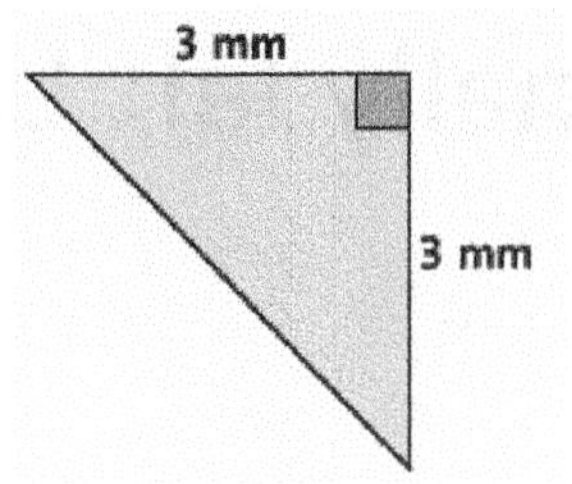

9. What is the length of the unknown side of the right triangle? Simplify your answer, if possible.

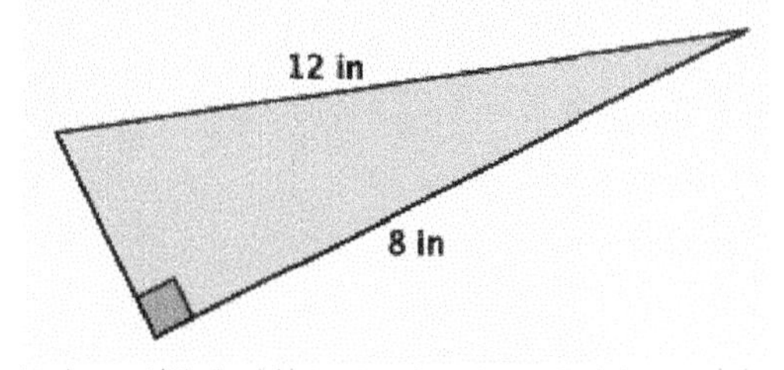

10. Josue simplified $\sqrt{450}$ as $15\sqrt{2}$. Is he correct? Explain why or why not.

11. Tiah was absent from school the day that you learned how to simplify a square root. Using $\sqrt{360}$, write Tiah an explanation for simplifying square roots.

This page intentionally left blank

Lesson 5: Solving Equations with Radicals

Classwork

Example 1

$$x^3 + 9x = \frac{1}{2}(18x + 54)$$

Example 2

$$x(x - 3) - 51 = -3x + 13$$

Exercises

Find the positive value of x that makes each equation true, and then verify your solution is correct.

1.

   a. Solve $x^2 - 14 = 5x + 67 - 5x$.

   b. Explain how you solved the equation.

2. Solve and simplify: $x(x-1) = 121 - x$

3. A square has a side length of $3x$ inches and an area of 324 in^2. What is the value of x?

4. $-3x^3 + 14 = -67$

5. $x(x + 4) - 3 = 4(x + 19.5)$

6. $216 + x = x(x^2 - 5) + 6x$

7.

a. What are we trying to determine in the diagram below?

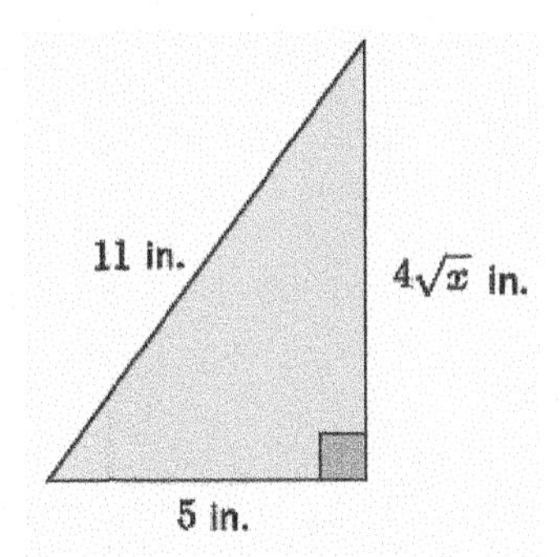

b. Determine the value of x, and check your answer.

Lesson Summary

Equations that contain variables that are squared or cubed can be solved using the properties of equality and the definition of square and cube roots.

Simplify an equation until it is in the form of $x^2 = p$ or $x^3 = p$, where p is a positive rational number; then, take the square or cube root to determine the positive value of x.

Example:

Solve for x.

$$\frac{1}{2}(2x^2 + 10) = 30$$
$$x^2 + 5 = 30$$
$$x^2 + 5 - 5 = 30 - 5$$
$$x^2 = 25$$
$$\sqrt{x^2} = \sqrt{25}$$
$$x = 5$$

Check:

$$\frac{1}{2}(2(5)^2 + 10) = 30$$
$$\frac{1}{2}(2(25) + 10) = 30$$
$$\frac{1}{2}(50 + 10) = 30$$
$$\frac{1}{2}(60) = 30$$
$$30 = 30$$

Problem Set

Find the positive value of x that makes each equation true, and then verify your solution is correct.

1. $x^2(x + 7) = \frac{1}{2}(14x^2 + 16)$

2. $x^3 = 1331^{-1}$

3. Determine the positive value of x that makes the equation true, and then explain how you solved the equation.
$$\frac{x^9}{x^7} - 49 = 0$$

4. Determine the positive value of x that makes the equation true.
$$(8x)^2 = 1$$

5. $(9\sqrt{x})^2 - 43x = 76$

6. Determine the length of the hypotenuse of the right triangle below.

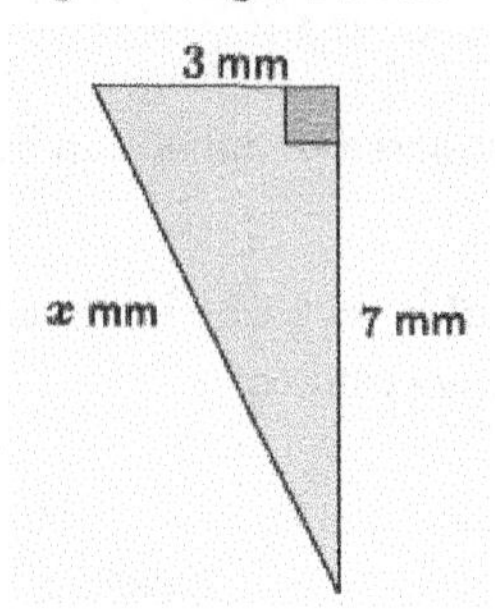

7. Determine the length of the legs in the right triangle below.

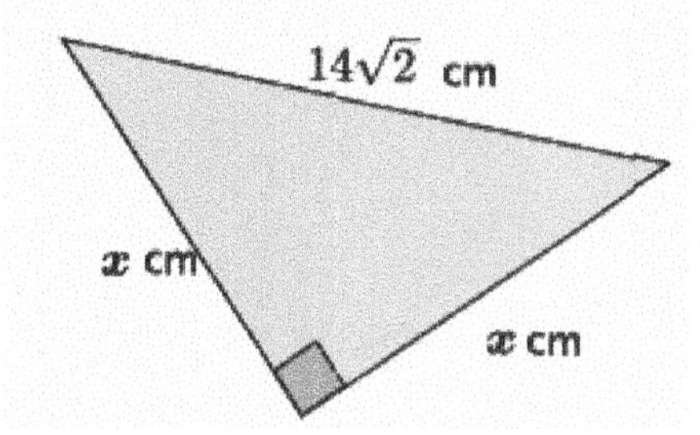

8. An equilateral triangle has side lengths of 6 cm. What is the height of the triangle? What is the area of the triangle?

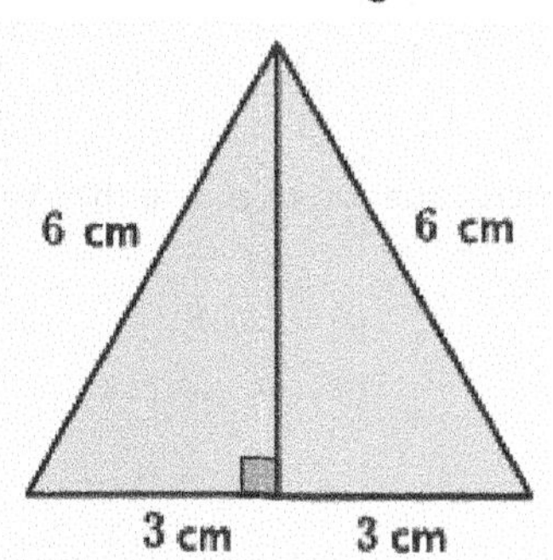

9. Challenge: Find the positive value of x that makes the equation true.

$$\left(\frac{1}{2}x\right)^2 - 3x = 7x + 8 - 10x$$

10. Challenge: Find the positive value of x that makes the equation true.

$$11x + x(x - 4) = 7(x + 9)$$

Lesson 6: Finite and Infinite Decimals

Classwork

Opening Exercise

a. Use long division to determine the decimal expansion of $\frac{54}{20}$.

b. Use long division to determine the decimal expansion of $\frac{7}{8}$.

c. Use long division to determine the decimal expansion of $\frac{8}{9}$.

d. Use long division to determine the decimal expansion of $\frac{22}{7}$.

e. What do you notice about the decimal expansions of parts (a) and (b) compared to the decimal expansions of parts (c) and (d)?

Example 1

Consider the fraction $\frac{5}{8}$. Write an equivalent form of this fraction with a denominator that is a power of 10, and hence write the decimal expansion of this fraction.

Example 2

Consider the fraction $\frac{17}{125}$. Is it equal to a finite or an infinite decimal? How do you know?

Exercises 1–5

You may use a calculator, but show your steps for each problem.

1. Consider the fraction $\frac{3}{8}$.

   a. Write the denominator as a product of 2's and/or 5's. Explain why this way of rewriting the denominator helps to find the decimal representation of $\frac{3}{8}$.

   b. Find the decimal representation of $\frac{3}{8}$. Explain why your answer is reasonable.

2. Find the first four places of the decimal expansion of the fraction $\frac{43}{64}$.

3. Find the first four places of the decimal expansion of the fraction $\frac{29}{125}$.

4. Find the first four decimal places of the decimal expansion of the fraction $\frac{19}{34}$.

5. Identify the type of decimal expansion for each of the numbers in Exercises 1–4 as finite or infinite. Explain why their decimal expansion is such.

Example 3

Will the decimal expansion of $\frac{7}{80}$ be finite or infinite? If it is finite, find it.

Example 4

Will the decimal expansion of $\frac{3}{160}$ be finite or infinite? If it is finite, find it.

Exercises 6–8

You may use a calculator, but show your steps for each problem.

6. Convert the fraction $\frac{37}{40}$ to a decimal.

   a. Write the denominator as a product of 2's and/or 5's. Explain why this way of rewriting the denominator helps to find the decimal representation of $\frac{37}{40}$.

   b. Find the decimal representation of $\frac{37}{40}$. Explain why your answer is reasonable.

7. Convert the fraction $\frac{3}{250}$ to a decimal.

8. Convert the fraction $\frac{7}{1250}$ to a decimal.

Lesson Summary

Fractions with denominators that can be expressed as products of 2's and/or 5's are equivalent to fractions with denominators that are a power of 10. These are precisely the fractions with finite decimal expansions.

Example:

Does the fraction $\frac{1}{8}$ have a finite or an infinite decimal expansion?

Since $8 = 2^3$, then the fraction has a finite decimal expansion. The decimal expansion is found as

$$\frac{1}{8} = \frac{1}{2^3} = \frac{1 \times 5^3}{2^3 \times 5^3} = \frac{125}{10^3} = 0.125.$$

If the denominator of a (simplified) fraction cannot be expressed as a product of 2's and/or 5's, then the decimal expansion of the number will be infinite.

Problem Set

Convert each fraction given to a finite decimal, if possible. If the fraction cannot be written as a finite decimal, then state how you know. You may use a calculator, but show your steps for each problem.

1. $\frac{2}{32}$
2. $\frac{99}{125}$
3. $\frac{15}{128}$
4. $\frac{8}{15}$
5. $\frac{3}{28}$
6. $\frac{13}{400}$
7. $\frac{5}{64}$
8. $\frac{15}{35}$
9. $\frac{199}{250}$
10. $\frac{219}{625}$

Lesson 7: Infinite Decimals

Classwork

Opening Exercise

a. Write the expanded form of the decimal 0.3765 using powers of 10.

b. Write the expanded form of the decimal $0.3333333...$ using powers of 10.

c. Have you ever wondered about the value of $0.99999...$? Some people say this infinite decimal has value 1. Some disagree. What do you think?

Example 1

The number 0.253 is represented on the number line below.

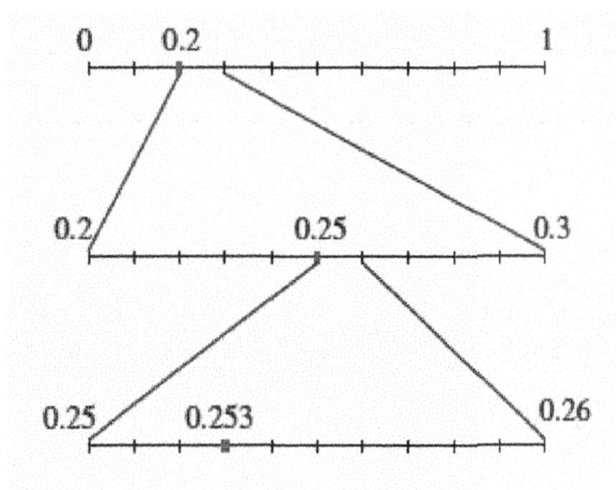

Example 2

The number $\frac{5}{6}$ which is equal to $0.833333...$ or $0.8\overline{3}$ is partially represented on the number line below.

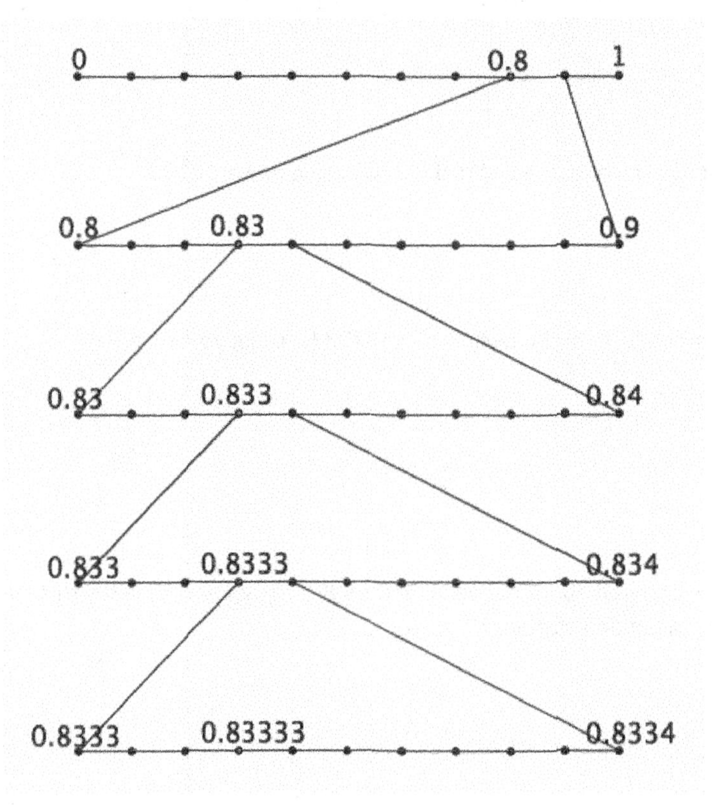

Exercises 1–5

1.

a. Write the expanded form of the decimal 0.125 using powers of 10.

b. Show on the number line the placement of the decimal 0.125.

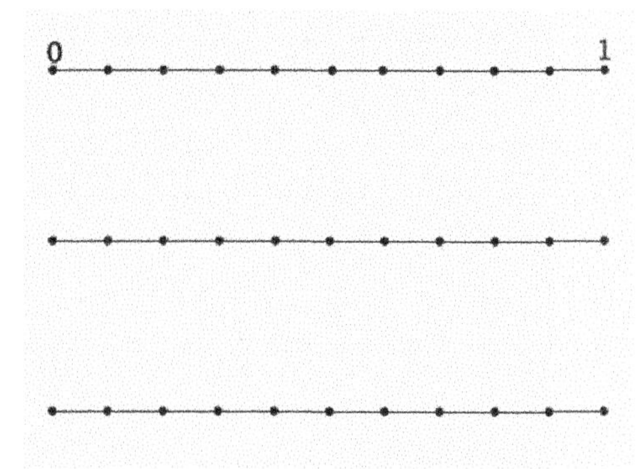

2.

a. Write the expanded form of the decimal 0.3875 using powers of 10.

b. Show on the number line the placement of the decimal 0.3875.

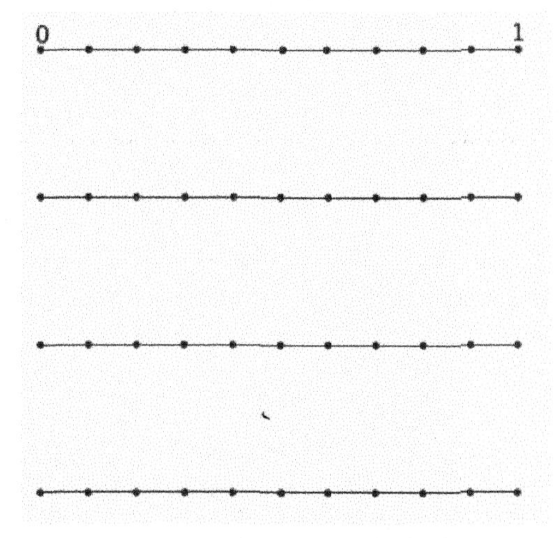

3.

a. Write the expanded form of the decimal $0.777777...$ using powers of 10.

b. Show the first few stages of placing the decimal $0.777777...$ on the number line.

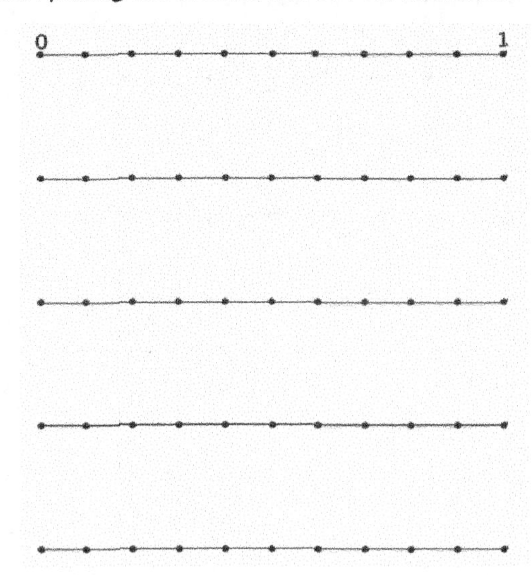

4.

a. Write the expanded form of the decimal $0.\overline{45}$ using powers of 10.

b. Show the first few stages of placing the decimal $0.\overline{45}$ on the number line.

5.

a. Order the following numbers from least to greatest: 2.121212, 2.1, 2.2, and $2.\overline{12}$.

b. Explain how you knew which order to put the numbers in.

Lesson Summary

An infinite decimal is a decimal whose expanded form is infinite.

Example:
The expanded form of the decimal $0.8\overline{3} = 0.83333...$ is $\frac{8}{10} + \frac{3}{10^2} + \frac{3}{10^3} + \frac{3}{10^4} + \cdots$.

To pin down the placement of an infinite decimal on the number line, we first identify within which tenth it lies, then within which hundredth it lies, then within which thousandth, and so on. These intervals have widths getting closer and closer to a width of zero.

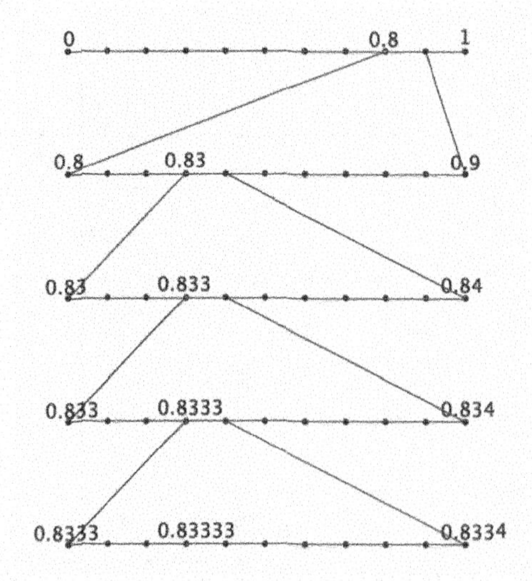

This reasoning allows us to deduce that the infinite decimal $0.9999...$ and 1 have the same location on the number line. Consequently, $0.\overline{9} = 1$.

Problem Set

1.

a. Write the expanded form of the decimal 0.625 using powers of 10.

b. Place the decimal 0.625 on the number line.

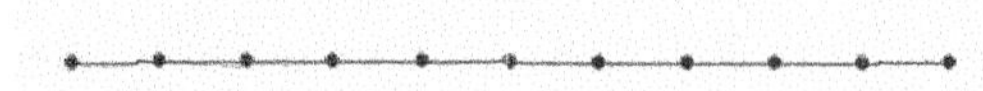

2.

a. Write the expanded form of the decimal $0.\overline{370}$ using powers of 10.

b. Show the first few stages of placing the decimal $0.370370\ldots$ on the number line.

3. Which is a more accurate representation of the fraction $\frac{2}{3}$: 0.6666 or $0.\overline{6}$? Explain. Which would you prefer to compute with?

4. Explain why we shorten infinite decimals to finite decimals to perform operations. Explain the effect of shortening an infinite decimal on our answers.

5. A classmate missed the discussion about why $0.\overline{9} = 1$. Convince your classmate that this equality is true.

6. Explain why $0.3333 < 0.33333$.

This page intentionally left blank

Lesson 8: The Long Division Algorithm

Classwork

Example 1

Show that the decimal expansion of $\frac{26}{4}$ is 6.5.

Exploratory Challenge/Exercises 1–5

1.

a. Use long division to determine the decimal expansion of $\frac{142}{2}$.

b. Fill in the blanks to show another way to determine the decimal expansion of $\frac{142}{2}$.

$$142 = \_\_\_ \times 2 + \_\_\_$$

$$\frac{142}{2} = \frac{\_\_\_ \times 2 + \_\_\_}{2}$$

$$\frac{142}{2} = \frac{\_\_\_ \times 2}{2} + \frac{\_\_\_}{2}$$

$$\frac{142}{2} = \_\_\_ + \frac{\_\_\_}{2}$$

$$\frac{142}{2} = \_\_\_\_$$

c. Does the number $\frac{142}{2}$ have a finite or an infinite decimal expansion?

2.

a. Use long division to determine the decimal expansion of $\frac{142}{4}$.

b. Fill in the blanks to show another way to determine the decimal expansion of $\frac{142}{4}$.

$$142 = \_\_\_\_ \times 4 + \_\_\_\_$$

$$\frac{142}{4} = \frac{\_\_\_\_ \times 4 + \_\_\_\_}{4}$$

$$\frac{142}{4} = \frac{\_\_\_\_ \times 4}{4} + \frac{\_\_\_\_}{4}$$

$$\frac{142}{4} = \_\_\_\_ + \frac{\_\_\_\_}{4}$$

$$\frac{142}{4} = \_\_\_\_\_\_$$

c. Does the number $\frac{142}{4}$ have a finite or an infinite decimal expansion?

3.

a. Use long division to determine the decimal expansion of $\frac{142}{6}$.

b. Fill in the blanks to show another way to determine the decimal expansion of $\frac{142}{6}$.

$$142 = \_\_\_\_ \times 6 + \_\_\_\_$$

$$\frac{142}{6} = \frac{\_\_\_\_ \times 6 + \_\_\_\_}{6}$$

$$\frac{142}{6} = \frac{\_\_\_\_ \times 6}{6} + \frac{\_\_\_\_}{6}$$

$$\frac{142}{6} = \_\_\_\_ + \frac{\_\_\_\_}{6}$$

$$\frac{142}{6} = \_\_\_\_\_\_\_\_$$

c. Does the number $\frac{142}{6}$ have a finite or an infinite decimal expansion?

4.

a. Use long division to determine the decimal expansion of $\frac{142}{11}$.

b. Fill in the blanks to show another way to determine the decimal expansion of $\frac{142}{11}$.

$$142 = \_\_\_\_ \times 11 + \_\_\_\_$$

$$\frac{142}{11} = \frac{\_\_\_\_ \times 11 + \_\_\_\_}{11}$$

$$\frac{142}{11} = \frac{\_\_\_\_ \times 11}{11} + \frac{\_\_\_\_}{11}$$

$$\frac{142}{11} = \_\_\_\_ + \frac{\_\_\_\_}{11}$$

$$\frac{142}{11} = \_\_\_\_\_\_\_\_$$

c. Does the number $\frac{142}{11}$ have a finite or an infinite decimal expansion?

5. In general, which fractions produce infinite decimal expansions?

Exercises 6–10

6. Does the number $\frac{65}{13}$ have a finite or an infinite decimal expansion? Does its decimal expansion have a repeating pattern?

7. Does the number $\frac{17}{11}$ have a finite or an infinite decimal expansion? Does its decimal expansion have a repeating pattern?

8. Is the number $0.212112111211112111112...$ rational? Explain. (Assume the pattern you see in the decimal expansion continues.)

9. Does the number $\frac{860}{999}$ have a finite or an infinite decimal expansion? Does its decimal expansion have a repeating pattern?

10. Is the number $0.1234567891011121314151617181920212223...$ rational? Explain. (Assume the pattern you see in the decimal expansion continues.)

Lesson Summary

A rational number is a number that can be written in the form $\frac{a}{b}$ for a pair of integers a and b with b not zero.

The long division algorithm shows that every rational number has a decimal expansion that falls into a repeating pattern. For example, the rational number 32 has a decimal expansion of $32.\bar{0}$, the rational number $\frac{1}{3}$ has a decimal expansion of $0.\bar{3}$, and the rational number $\frac{4}{11}$ has a decimal expansion of $0.\overline{45}$.

Problem Set

1. Write the decimal expansion of $\frac{7000}{9}$ as an infinitely long repeating decimal.

2. Write the decimal expansion of $\frac{6555555}{3}$ as an infinitely long repeating decimal.

3. Write the decimal expansion of $\frac{350000}{11}$ as an infinitely long repeating decimal.

4. Write the decimal expansion of $\frac{1200000}{37}$ as an infinitely long repeating decimal.

5. Someone notices that the long division of 2,222,222 by 6 has a quotient of $370{,}370$ and a remainder of 2 and wonders why there is a repeating block of digits in the quotient, namely 370. Explain to the person why this happens.

6. Is the answer to the division problem number $10 \div 3.2$ a rational number? Explain.

7. Is $\frac{3\pi}{77\pi}$ a rational number? Explain.

8. The decimal expansion of a real number x has every digit 0 except the first digit, the tenth digit, the hundredth digit, the thousandth digit, and so on, are each 1. Is x a rational number? Explain.

This page intentionally left blank

Lesson 9: Decimal Expansions of Fractions, Part 1

Classwork

Opening Exercise

a. Compute the decimal expansions of $\frac{5}{6}$ and $\frac{7}{9}$.

b. What is $\frac{5}{6}+\frac{7}{9}$ as a fraction? What is the decimal expansion of this fraction?

c. What is $\frac{5}{6}\times\frac{7}{9}$ as a fraction? According to a calculator, what is the decimal expansion of the answer?

d. If you were given just the decimal expansions of $\frac{5}{6}$ and $\frac{7}{9}$, without knowing which fractions produced them, do you think you could easily add the two decimals to find the decimal expansion of their sum? Could you easily multiply the two decimals to find the decimal expansion of their product?

Exercise 1

Two irrational numbers x and y have infinite decimal expansions that begin $0.67035267\ldots$ for x and $0.84991341\ldots$ for y.

a. Explain why 0.670 is an approximation for x with an error of less than one thousandth. Explain why 0.849 is an approximation for y with an error of less than one thousandth.

b. Using the approximations given in part (a), what is an approximate value for $x + y$, for $x \times y$, and for $x^2 + 7y^2$?

c. Repeat part (b), but use approximations for x and y that have errors less than $\frac{1}{10^5}$.

Exercise 2

Two real numbers have decimal expansions that begin with the following:

$$x = 0.1538461....$$

$$y = 0.3076923....$$

a. Using approximations for x and y that are accurate within a measure of $\frac{1}{10^3}$, find approximate values for $x + y$ and $y - 2x$.

b. Using approximations for x and y that are accurate within a measure of $\frac{1}{10^7}$, find approximate values for $x + y$ and $y - 2x$.

c. We now reveal that $x = \frac{2}{13}$ and $y = \frac{4}{13}$. How accurate is your approximate value to $y - 2x$ from part (a)? From part (b)?

d. Compute the first seven decimal places of $\frac{6}{13}$. How accurate is your approximate value to $x + y$ from part (a)? From part (b)?

Lesson Summary

It is not clear how to perform arithmetic on numbers given as infinitely long decimals. If we approximate these numbers by truncating their infinitely long decimal expansions to a finite number of decimal places, then we can perform arithmetic on the approximate values to estimate answers.

Truncating a decimal expansion to n decimal places gives an approximation with an error of less than $\frac{1}{10^n}$. For example, 0.676 is an approximation for $0.676767...$ with an error of less than 0.001.

Problem Set

1. Two irrational numbers x and y have infinite decimal expansions that begin $0.3338117\ ...$ for x and $0.9769112...$ for y.
   a. Explain why 0.33 is an approximation for x with an error of less than one hundredth. Explain why 0.97 is an approximation for y with an error of less than one hundredth.
   b. Using the approximations given in part (a), what is an approximate value for $2x(y+1)$?
   c. Repeat part (b), but use approximations for x and y that have errors less than $\frac{1}{10^6}$.

2. Two real numbers have decimal expansions that begin with the following:

$$x = 0.70588....$$
$$y = 0.23529....$$

   a. Using approximations for x and y that are accurate within a measure of $\frac{1}{10^2}$, find approximate values for $x + 1.25y$ and $\frac{x}{y}$.
   b. Using approximations for x and y that are accurate within a measure of $\frac{1}{10^4}$, find approximate values for $x + 1.25y$ and $\frac{x}{y}$.
   c. We now reveal that x and y are rational numbers with the property that each of the values $x + 1.25y$ and $\frac{x}{y}$ is a whole number. Make a guess as to what whole numbers these values are, and use your guesses to find what fractions x and y might be.

Lesson 10: Converting Repeating Decimals to Fractions

Classwork

Example 1

There is a fraction with an infinite decimal expansion of $0.\overline{81}$. Find the fraction.

Exercises 1–2

1. There is a fraction with an infinite decimal expansion of $0.\overline{123}$. Let $x = 0.\overline{123}$.
   a. Explain why looking at $1000x$ helps us find the fractional representation of x.

b. What is x as a fraction?

c. Is your answer reasonable? Check your answer using a calculator.

2. There is a fraction with a decimal expansion of $0.\overline{4}$. Find the fraction, and check your answer using a calculator.

Example 2

Could it be that $2.13\overline{8}$ is also a fraction?

Exercises 3–4

3. Find the fraction equal to $1.6\overline{23}$. Check your answer using a calculator.

4. Find the fraction equal to $2.9\overline{60}$. Check your answer using a calculator.

Lesson Summary

Every decimal with a repeating pattern is a rational number, and we have the means to determine the fraction that has a given repeating decimal expansion.

Example: Find the fraction that is equal to the number $0.\overline{567}$.

Let x represent the infinite decimal $0.\overline{567}$.

$x = 0.\overline{567}$	
$10^3x = 10^3(0.\overline{567})$	Multiply by 10^3 because there are 3 digits that repeat.
$1000x = 567.\overline{567}$	Simplify
$1000x = 567 + 0.\overline{567}$	By addition
$1000x = 567 + x$	By substitution; $x = 0.\overline{567}$
$1000x - x = 567 + x - x$	Subtraction property of equality
$999x = 567$	Simplify
$\frac{999}{999}x = \frac{567}{999}$	Division property of equality
$x = \frac{567}{999} = \frac{63}{111}$	Simplify

This process may need to be used more than once when the repeating digits, as in numbers such as $1.2\overline{6}$, do not begin immediately after the decimal.

Irrational numbers are numbers that are not rational. They have infinite decimal expansions that do not repeat and they cannot be expressed as $\frac{p}{q}$ for integers p and q with $q \neq 0$.

Problem Set

1.
   a. Let $x = 0.\overline{631}$. Explain why multiplying both sides of this equation by 10^3 will help us determine the fractional representation of x.
   b. What fraction is x?
   c. Is your answer reasonable? Check your answer using a calculator.

2. Find the fraction equal to $3.40\overline{8}$. Check your answer using a calculator.

3. Find the fraction equal to $0.\overline{5923}$. Check your answer using a calculator.

4. Find the fraction equal to $2.3\overline{82}$. Check your answer using a calculator.

5. Find the fraction equal to $0.\overline{714285}$. Check your answer using a calculator.

6. Explain why an infinite decimal that is not a repeating decimal cannot be rational.

7. In a previous lesson, we were convinced that it is acceptable to write $0.\overline{9} = 1$. Use what you learned today to show that it is true.

8. Examine the following repeating infinite decimals and their fraction equivalents. What do you notice? Why do you think what you observed is true?

$$0.\overline{81} = \frac{81}{99} \qquad 0.\overline{4} = \frac{4}{9} \qquad 0.\overline{123} = \frac{123}{999} \qquad 0.\overline{60} = \frac{60}{99}$$

$$0.\overline{4311} = \frac{4311}{9999} \qquad 0.\overline{01} = \frac{1}{99} \qquad 0.\overline{3} = \frac{1}{3} = \frac{3}{9} \qquad 0.\overline{9} = 1.0$$

Lesson 11: The Decimal Expansion of Some Irrational Numbers

Classwork

Opening Exercise

Place $\sqrt{28}$ on a number line. Make a guess as to the first few values of the decimal expansion of $\sqrt{28}$. Explain your reasoning.

Example 1

Consider the decimal expansion of $\sqrt{3}$.

Find the first two values of the decimal expansion using the following fact: If $c^2 < 3 < d^2$ for positive numbers c and d, then $c < \sqrt{3} < d$.

First Approximation:

Because $1 < 3 < 4$, we have $1 < \sqrt{3} < 2$.

Second approximation:

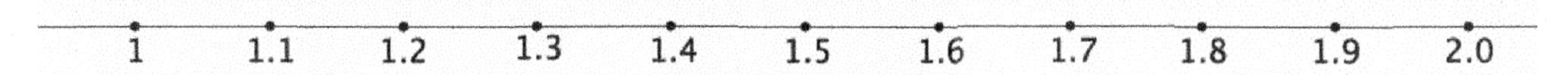

Third approximation:

Example 2

Find the first few places of the decimal expansion of $\sqrt{28}$.

First approximation:

Second approximation:

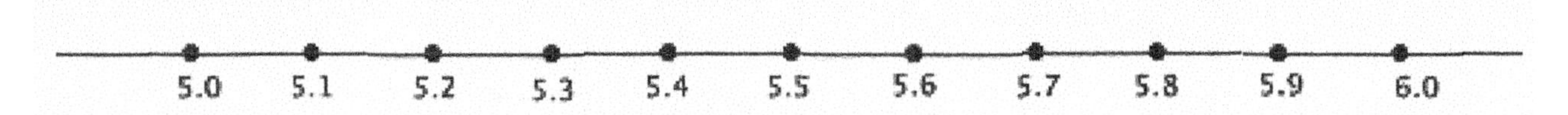

Third approximation:

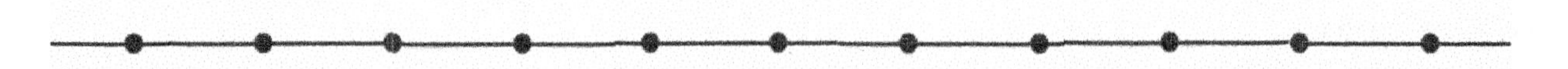

Fourth approximation:

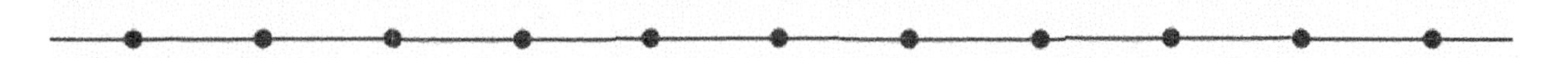

Exercise 1

In which interval of hundredths does $\sqrt{14}$ lie? Show your work.

Lesson Summary

To find the first few decimal places of the decimal expansion of the square root of a non-perfect square, first determine between which two integers the square root lies, then in which interval of a tenth the square root lies, then in which interval of a hundredth it lies, and so on.

Example: Find the first few decimal places of $\sqrt{22}$.

Begin by determining between which two integers the number would lie.
$\sqrt{22}$ is between the integers 4 and 5 because $16 < 22 < 25$.

Next, determine between which interval of tenths the number belongs.
$\sqrt{22}$ is between 4.6 and 4.7 because $4.6^2 = 21.16 < 22 < 4.7^2 = 22.09$.

Next, determine between which interval of hundredths the number belongs.
$\sqrt{22}$ is between 4.69 and 4.70 because $4.69^2 = 21.9961 < 22 < 4.70^2 = 22.0900$.

A good estimate of the value of $\sqrt{22}$ is 4.69. It is correct to two decimal places and so has an error no larger than 0.01.

Notice that with each step of this process we are getting closer and closer to the actual value $\sqrt{22}$. This process can continue using intervals of thousandths, ten-thousandths, and so on.

Problem Set

1. In which hundredth interval of the number line does $\sqrt{84}$ lie?

2. Determine the three-decimal digit approximation of the number $\sqrt{34}$.

3. Write the decimal expansion of $\sqrt{47}$ to at least two-decimal digits.

4. Write the decimal expansion of $\sqrt{46}$ to at least two-decimal digits.

5. Explain how to improve the accuracy of the decimal expansion of an irrational number.

6. Is the number $\sqrt{144}$ rational or irrational? Explain.

7. Is the number $0.\overline{64} = 0.646464646\ldots$ rational or irrational? Explain.

8. Henri computed the first 100 decimal digits of the number $\frac{352}{541}$ and got

$$0.6506469500924214417744916820702402957486136783733826247689463955637707948243992606284658040665434380776340110905730129 4\ldots$$

He saw no repeating pattern to the decimal and so concluded that the number is irrational. Do you agree with Henri's conclusion? If not, what would you say to Henri?

9. Use a calculator to determine the decimal expansion of $\sqrt{35}$. Does the number appear to be rational or irrational? Explain.

10. Use a calculator to determine the decimal expansion of $\sqrt{101}$. Does the number appear to be rational or irrational? Explain.

11. Use a calculator to determine the decimal expansion of $\sqrt{7}$. Does the number appear to be rational or irrational? Explain.

12. Use a calculator to determine the decimal expansion of $\sqrt{8720}$. Does the number appear to be rational or irrational? Explain.

13. Use a calculator to determine the decimal expansion of $\sqrt{17956}$. Does the number appear to be rational or irrational? Explain.

14. Since the number $\frac{3}{5}$ is rational, must the number $\left(\frac{3}{5}\right)^2$ be rational as well? Explain.

15. If a number x is rational, must the number x^2 be rational as well? Explain.

16. Challenge: Determine the two-decimal digit approximation of the number $\sqrt[3]{9}$.

This page intentionally left blank

Lesson 12: Decimal Expansions of Fractions, Part 2

Classwork

Example 1

Find the decimal expansion of $\frac{35}{11}$.

Exercises 1–3

1. Find the decimal expansion of $\frac{5}{3}$ without using long division.

2. Find the decimal expansion of $\frac{5}{11}$ without using long division.

3. Find the decimal expansion of the number $\frac{23}{99}$ first without using long division and then again using long division.

Lesson Summary

For rational numbers, there is no need to guess and check in which interval of tenths, hundredths, or thousandths the number will lie.

For example, to determine where the fraction $\frac{1}{8}$ lies in the interval of tenths, compute using the following inequality:

$\frac{m}{10} < \frac{1}{8} < \frac{m+1}{10}$	Use the denominator of 10 because we need to find the tenths digit of $\frac{1}{8}$.
$m < \frac{10}{8} < m+1$	Multiply through by 10.
$m < 1\frac{1}{4} < m+1$	Simplify the fraction $\frac{10}{8}$.

The last inequality implies that $m = 1$ and $m + 1 = 2$ because $1 < 1\frac{1}{4} < 2$. Then, the tenths digit of the decimal expansion of $\frac{1}{8}$ is 1.

To find in which interval of hundredths $\frac{1}{8}$ lies, we seek consecutive integers m and $m + 1$ so that

$$\frac{1}{10} + \frac{m}{100} < \frac{1}{8} < \frac{1}{10} + \frac{m+1}{100}.$$

This is equivalent to

$$\frac{m}{100} < \frac{1}{8} - \frac{1}{10} < \frac{m+1}{100},$$

so we compute $\frac{1}{8} - \frac{1}{10} = \frac{2}{80} = \frac{1}{40}$. We have

$$\frac{m}{100} < \frac{1}{40} < \frac{m+1}{100}.$$

Multiplying through by 100 gives

$$m < \frac{10}{4} < m + 1.$$

The last inequality implies that $m = 2$ and $m + 1 = 3$ because $2 < 2\frac{1}{2} < 3$. Then, the hundredths digit of the decimal expansion of $\frac{1}{8}$ is 2.

We can continue the process until the decimal expansion is complete or until we suspect a repeating pattern that we can verify.

Problem Set

1. Without using long division, explain why the tenths digit of $\frac{3}{11}$ is a 2.

2. Find the decimal expansion of $\frac{25}{9}$ without using long division.

3. Find the decimal expansion of $\frac{11}{41}$ to at least 5 digits without using long division.

4. Which number is larger, $\sqrt{10}$ or $\frac{28}{9}$? Answer this question without using long division.

5. Sam says that $\frac{7}{11} = 0.63$, and Jaylen says that $\frac{7}{11} = 0.636$. Who is correct? Why?

Lesson 13: Comparing Irrational Numbers

Classwork

Exploratory Challenge/Exercises 1–11

1. Rodney thinks that $\sqrt[3]{64}$ is greater than $\frac{17}{4}$. Sam thinks that $\frac{17}{4}$ is greater. Who is right and why?

2. Which number is smaller, $\sqrt[3]{27}$ or 2.89? Explain.

3. Which number is smaller, $\sqrt{121}$ or $\sqrt[3]{125}$? Explain.

4. Which number is smaller, $\sqrt{49}$ or $\sqrt[3]{216}$? Explain.

5. Which number is greater, $\sqrt{50}$ or $\frac{319}{45}$? Explain.

6. Which number is greater, $\frac{5}{11}$ or $0.\overline{4}$? Explain.

7. Which number is greater, $\sqrt{38}$ or $\frac{154}{25}$? Explain.

8. Which number is greater, $\sqrt{2}$ or $\frac{15}{9}$? Explain.

9. Place each of the following numbers at its approximate location on the number line: $\sqrt{25}$, $\sqrt{28}$, $\sqrt{30}$, $\sqrt{32}$, $\sqrt{35}$, and $\sqrt{36}$.

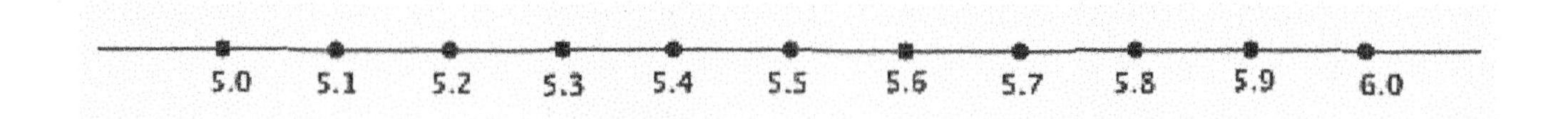

10. Challenge: Which number is larger, $\sqrt{5}$ or $\sqrt[3]{11}$?

11. A certain chessboard is being designed so that each square has an area of 3 in^2. What is the length of one edge of the board rounded to the tenths place? (A chessboard is composed of 64 squares as shown.)

Lesson Summary

Finding the first few places of the decimal expansion of numbers allows us to compare the numbers.

Problem Set

1. Which number is smaller, $\sqrt[3]{343}$ or $\sqrt{48}$? Explain.

2. Which number is smaller, $\sqrt{100}$ or $\sqrt[3]{1000}$? Explain.

3. Which number is larger, $\sqrt{87}$ or $\frac{929}{99}$? Explain.

4. Which number is larger, $\frac{9}{13}$ or $0.\overline{692}$? Explain.

5. Which number is larger, 9.1 or $\sqrt{82}$? Explain.

6. Place each of the following numbers at its approximate location on the number line: $\sqrt{144}$, $\sqrt[3]{1000}$, $\sqrt{130}$, $\sqrt{110}$, $\sqrt{120}$, $\sqrt{115}$, and $\sqrt{133}$. Explain how you knew where to place the numbers.

7. Which of the two right triangles shown below, measured in units, has the longer hypotenuse? Approximately how much longer is it?

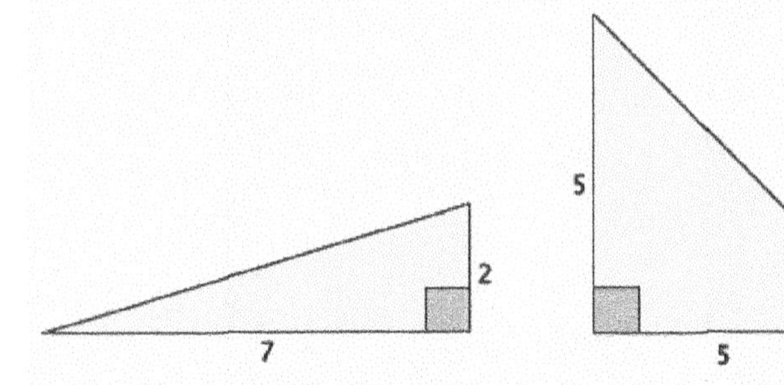

Lesson 14: Decimal Expansion of π

Classwork

Opening Exercise

a. Write an equation for the area, A, of the circle shown.

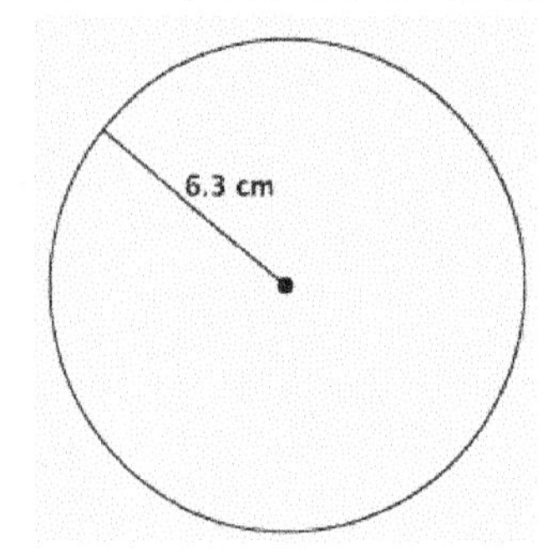

b. Write an equation for the circumference, C, of the circle shown.

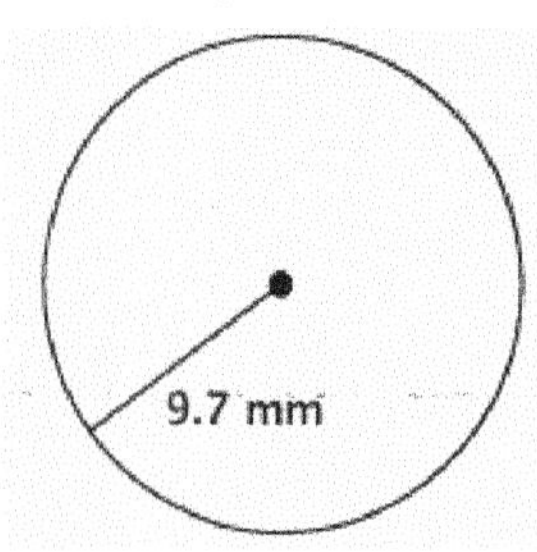

c. Each of the squares in the grid on the following page has an area of 1 unit^2.

i. Estimate the area of the circle shown by counting squares.

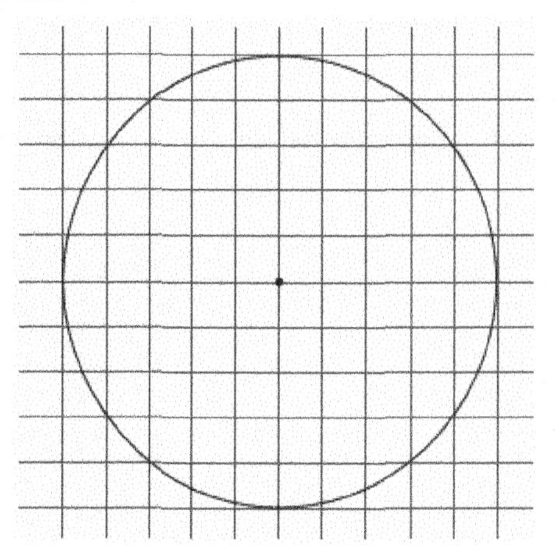

ii. Calculate the area of the circle using a radius of 5 units. Use 3.14 as an approximation for π.

Exercises 1–4

1. Gerald and Sarah are building a wheel with a radius of 6.5 cm and are trying to determine the circumference. Gerald says, "Because $6.5 \times 2 \times 3.14 = 40.82$, the circumference is 40.82 cm." Sarah says, "Because $6.5 \times 2 \times 3.10 = 40.3$ and $6.5 \times 2 \times 3.21 = 41.73$, the circumference is somewhere between 40.3 and 41.73." Explain the thinking of each student.

2. Estimate the value of the number $(6.12486\ldots)^2$.

3. Estimate the value of the number $(9.204107\ldots)^2$.

4. Estimate the value of the number $(4.014325\ldots)^2$.

Lesson Summary

Numbers, such as π, are frequently approximated in order to compute with them. Common approximations for π are 3.14 and $\frac{22}{7}$. It should be understood that using an approximate value of a number for computations produces an answer that is accurate to only the first few decimal digits.

Problem Set

1. Caitlin estimated π to be $3.10 < \pi < 3.21$. If she uses this approximation of π to determine the area of a circle with a radius of 5 cm, what could the area be?

2. Myka estimated the circumference of a circle with a radius of 4.5 in. to be 28.44 in. What approximate value of π did she use? Is it an acceptable approximation of π? Explain.

3. A length of ribbon is being cut to decorate a cylindrical cookie jar. The ribbon must be cut to a length that stretches the length of the circumference of the jar. There is only enough ribbon to make one cut. When approximating π to calculate the circumference of the jar, which number in the interval $3.10 < \pi < 3.21$ should be used? Explain.

4. Estimate the value of the number $(1.86211...)^2$.

5. Estimate the value of the number $(5.9035687...)^2$.

6. Estimate the value of the number $(12.30791...)^2$.

7. Estimate the value of the number $(0.6289731...)^2$.

8. Estimate the value of the number $(1.112223333...)^2$.

9. Which number is a better estimate for π, $\frac{22}{7}$ or 3.14? Explain.

10. To how many decimal digits can you correctly estimate the value of the number $(4.56789012...)^2$?

This page intentionally left blank

10 by 10 Grid

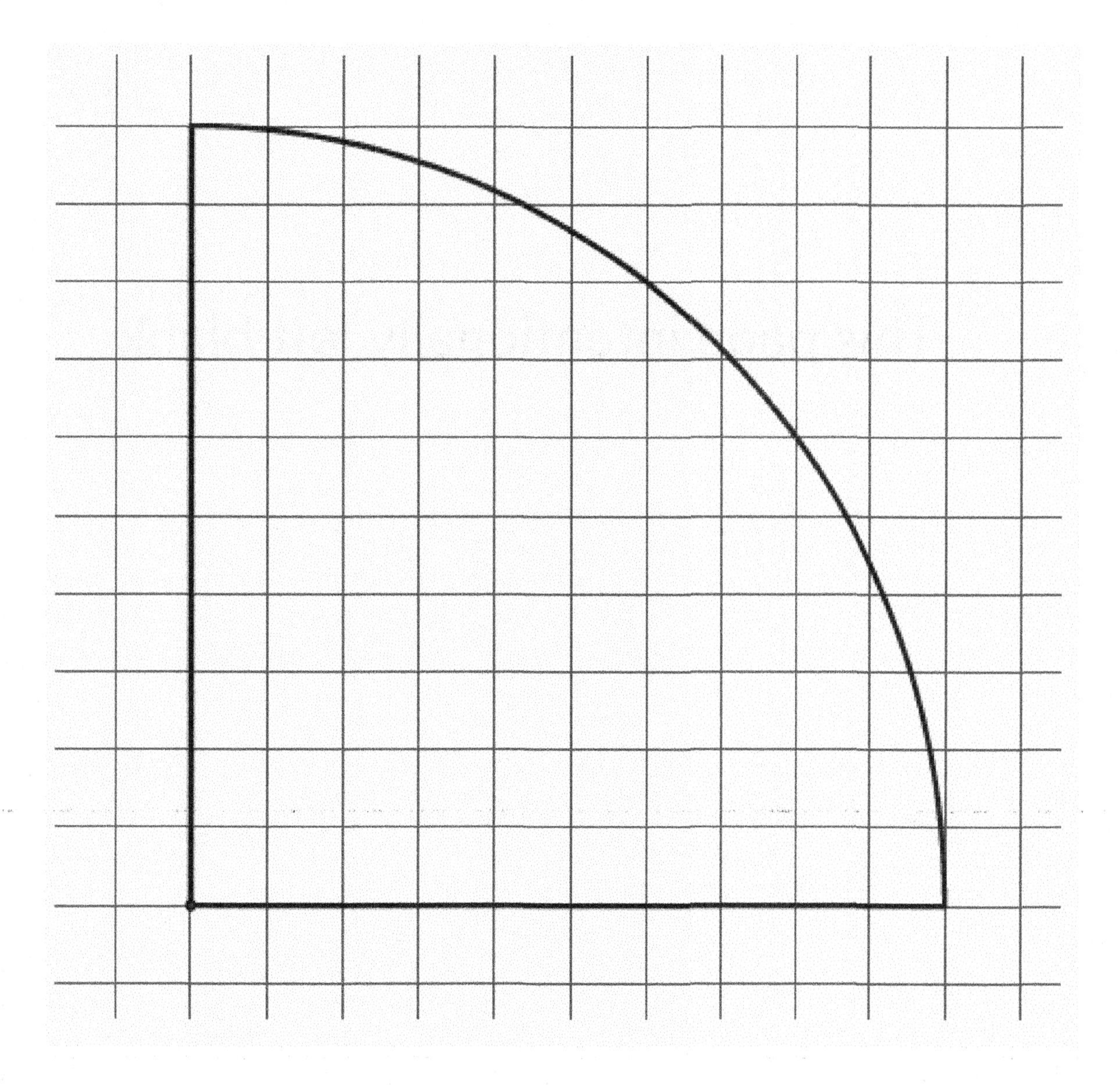

This page intentionally left blank

20 by 20 Grid

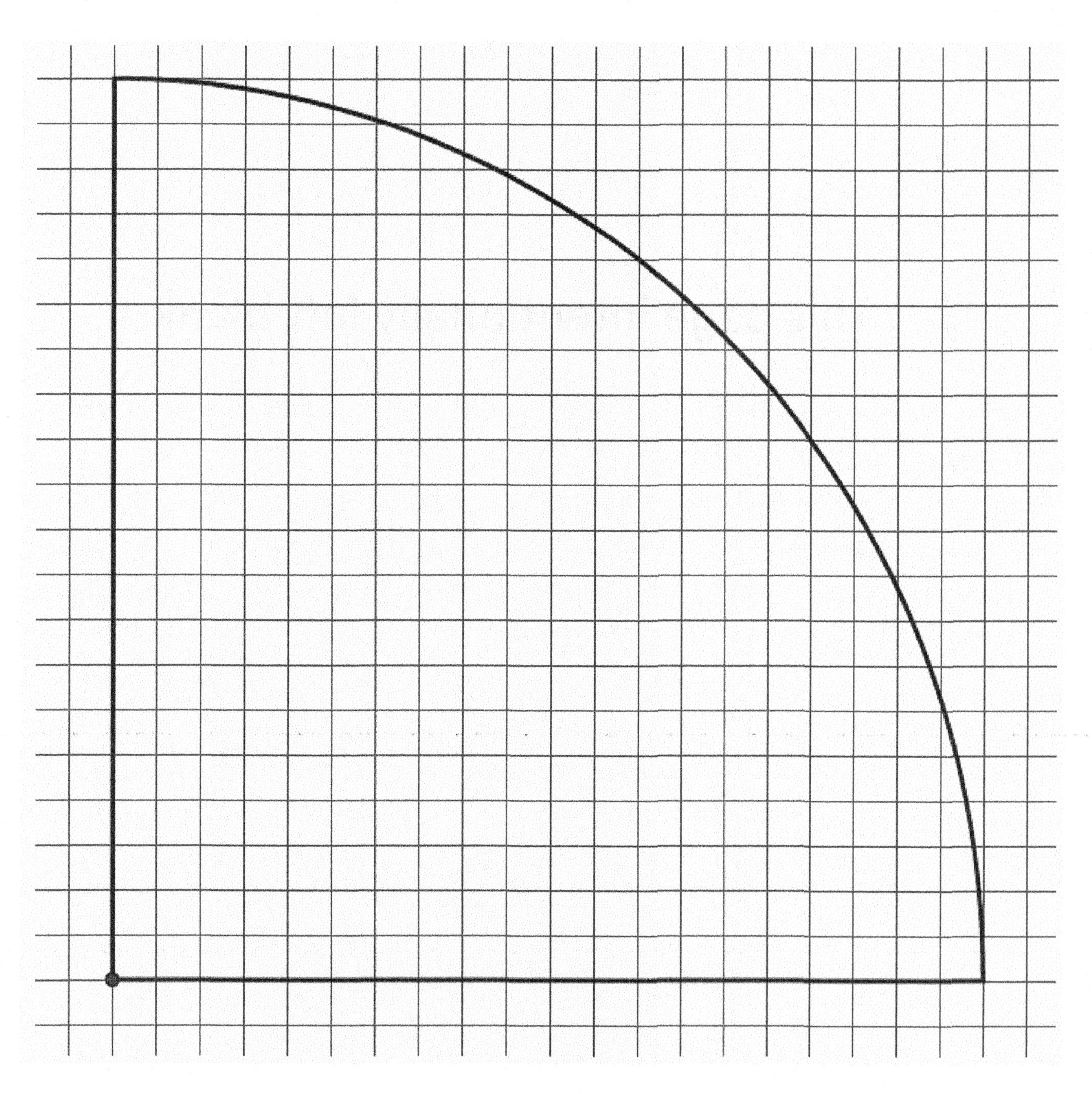

This page intentionally left blank

Lesson 15: Pythagorean Theorem, Revisited

Classwork

Proof of the Pythagorean Theorem

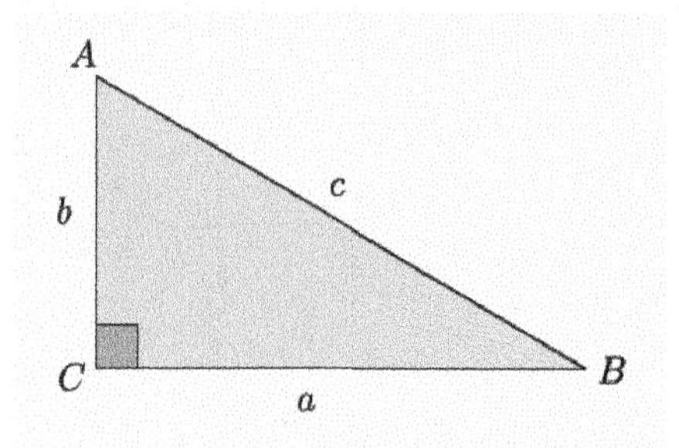

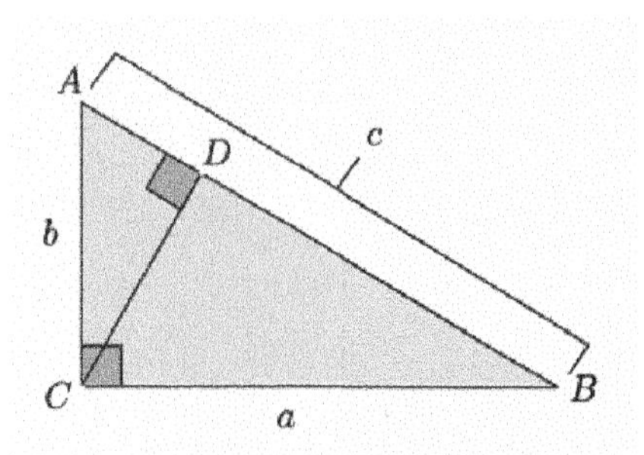

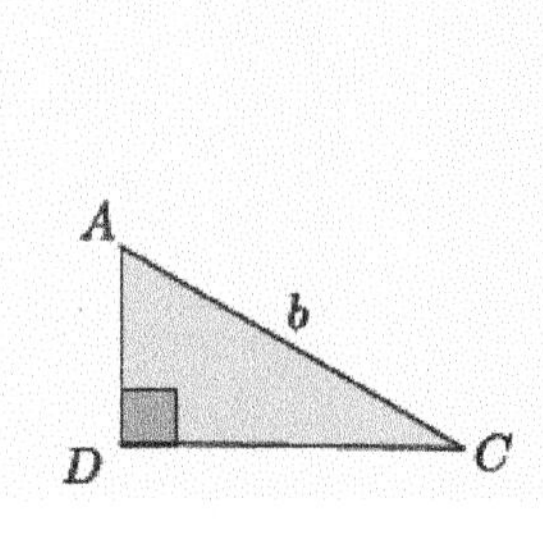

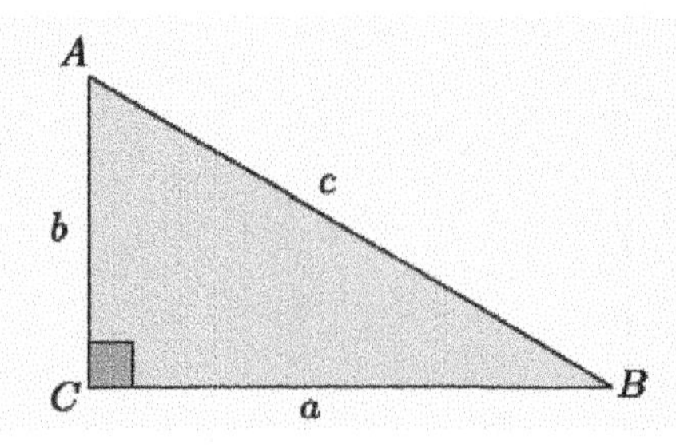

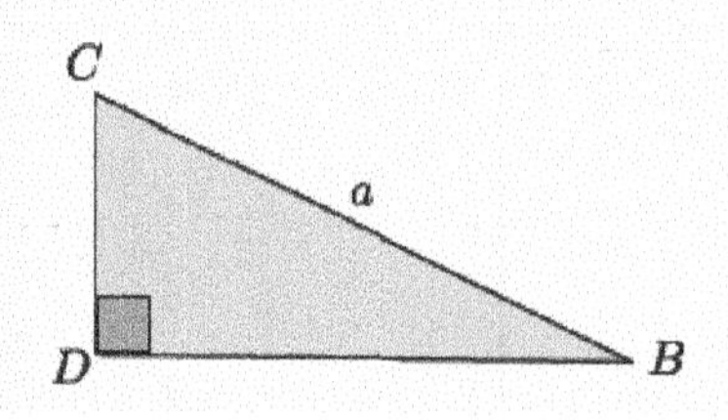

Discussion

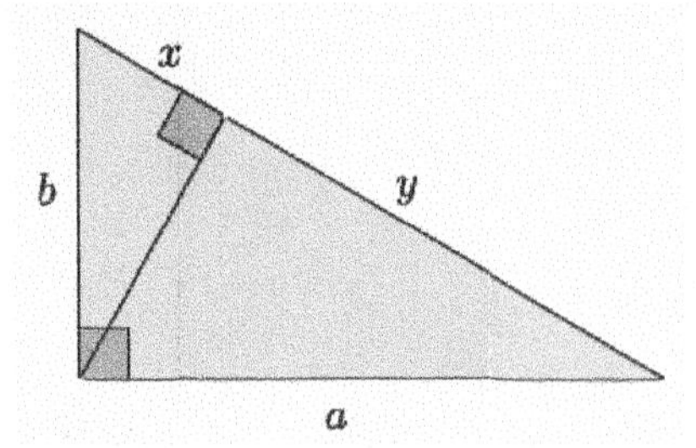

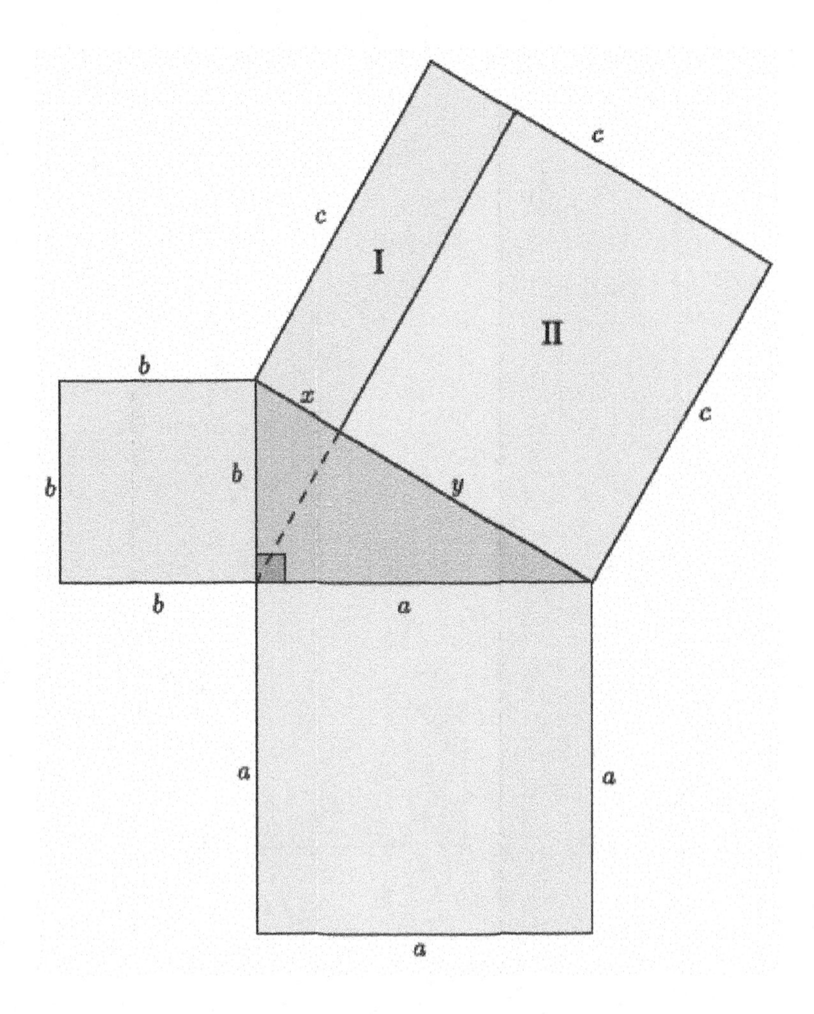

Lesson Summary

The Pythagorean theorem can be proven by drawing an altitude in the given right triangle and identifying three similar triangles. We can see geometrically how the large square drawn on the hypotenuse of the triangle has an area summing to the areas of the two smaller squares drawn on the legs of the right triangle.

Problem Set

1. For the right triangle shown below, identify and use similar triangles to illustrate the Pythagorean theorem.

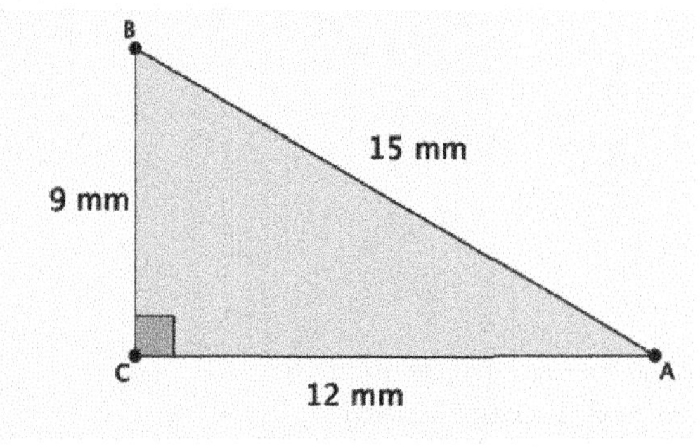

2. For the right triangle shown below, identify and use squares formed by the sides of the triangle to illustrate the Pythagorean theorem.

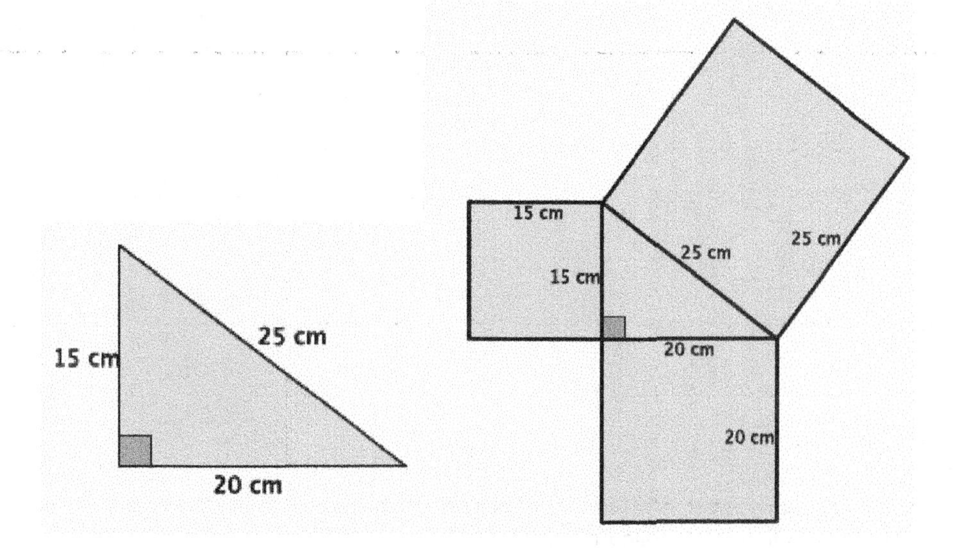

3. Reese claimed that any figure can be drawn off the sides of a right triangle and that as long as they are similar figures, then the sum of the areas off of the legs will equal the area off of the hypotenuse. She drew the diagram by constructing rectangles off of each side of a known right triangle. Is Reese's claim correct for this example? In order to prove or disprove Reese's claim, you must first show that the rectangles are similar. If they are, then you can use computations to show that the sum of the areas of the figures off of the sides a and b equals the area of the figure off of side c.

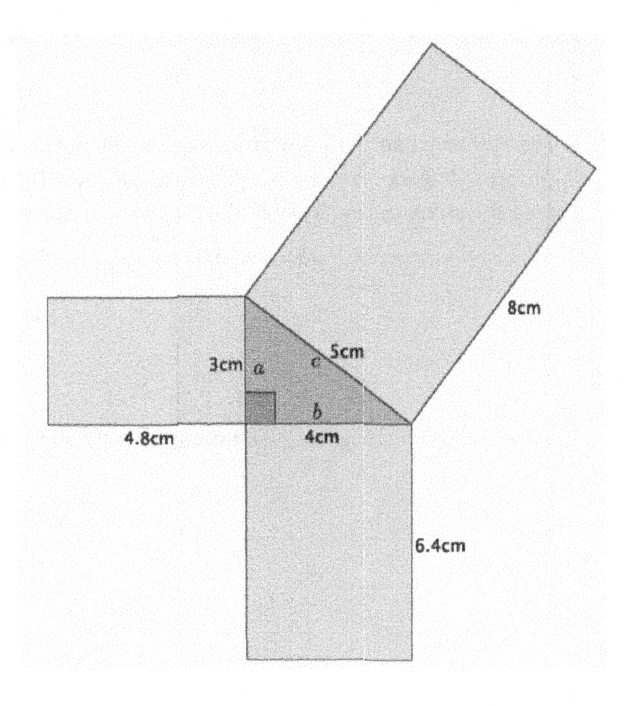

4. After learning the proof of the Pythagorean theorem using areas of squares, Joseph got really excited and tried explaining it to his younger brother. He realized during his explanation that he had done something wrong. Help Joseph find his error. Explain what he did wrong.

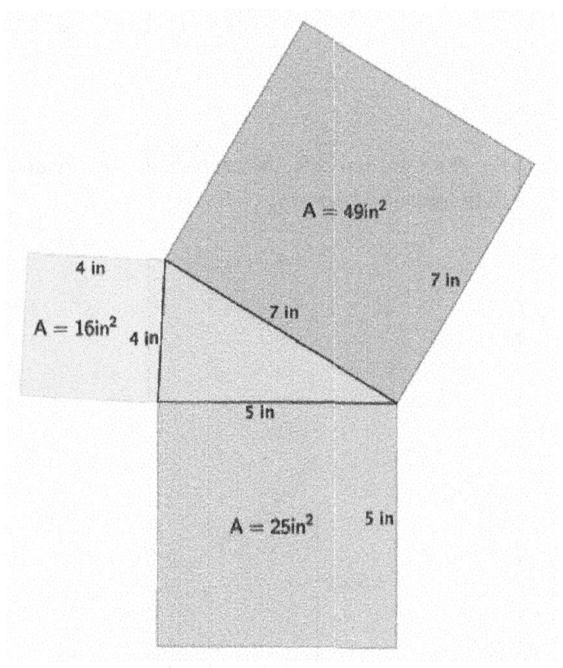

5. Draw a right triangle with squares constructed off of each side that Joseph can use the next time he wants to show his younger brother the proof of the Pythagorean theorem.

6. Explain the meaning of the Pythagorean theorem in your own words.

7. Draw a diagram that shows an example illustrating the Pythagorean theorem.

Lesson 16: Converse of the Pythagorean Theorem

Classwork

Proof of the Converse of the Pythagorean Theorem

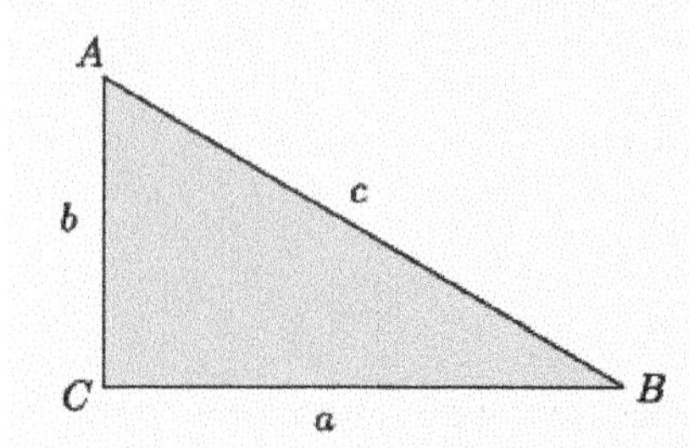

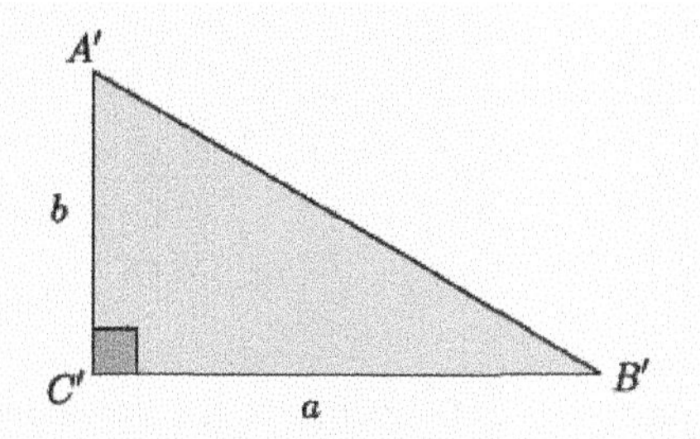

Exercises 1–7

1. Is the triangle with leg lengths of 3 mi. and 8 mi. and hypotenuse of length $\sqrt{73}$ mi. a right triangle? Show your work, and answer in a complete sentence.

2. What is the length of the unknown side of the right triangle shown below? Show your work, and answer in a complete sentence. Provide an exact answer and an approximate answer rounded to the tenths place.

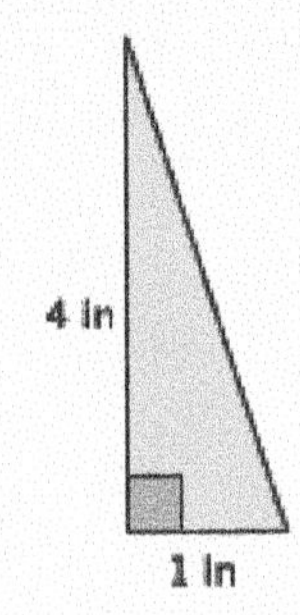

3. What is the length of the unknown side of the right triangle shown below? Show your work, and answer in a complete sentence. Provide an exact answer and an approximate answer rounded to the tenths place.

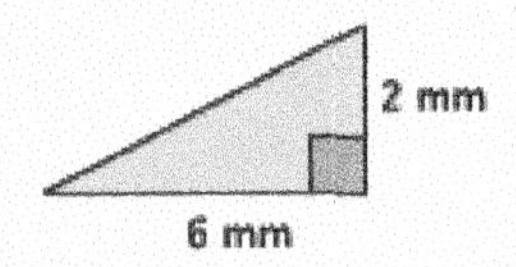

4. Is the triangle with leg lengths of 9 in. and 9 in. and hypotenuse of length $\sqrt{175}$ in. a right triangle? Show your work, and answer in a complete sentence.

5. Is the triangle with leg lengths of $\sqrt{28}$ cm and 6 cm and hypotenuse of length 8 cm a right triangle? Show your work, and answer in a complete sentence.

6. What is the length of the unknown side of the right triangle shown below? Show your work, and answer in a complete sentence.

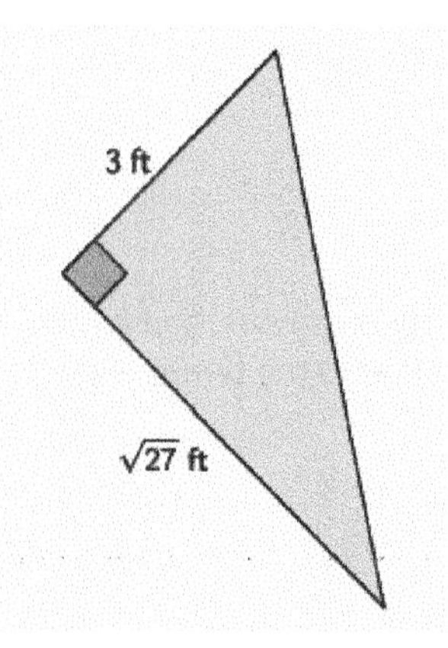

7. The triangle shown below is an isosceles right triangle. Determine the length of the legs of the triangle. Show your work, and answer in a complete sentence.

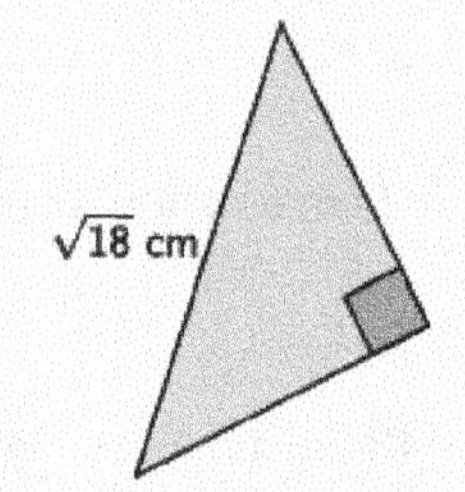

Lesson Summary

The converse of the Pythagorean theorem states that if a triangle with side lengths a, b, and c satisfies $a^2 + b^2 = c^2$, then the triangle is a right triangle.

The converse can be proven using concepts related to congruence.

Problem Set

1. What is the length of the unknown side of the right triangle shown below? Show your work, and answer in a complete sentence. Provide an exact answer and an approximate answer rounded to the tenths place.

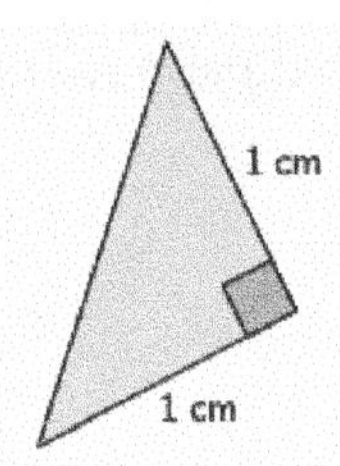

2. What is the length of the unknown side of the right triangle shown below? Show your work, and answer in a complete sentence. Provide an exact answer and an approximate answer rounded to the tenths place.

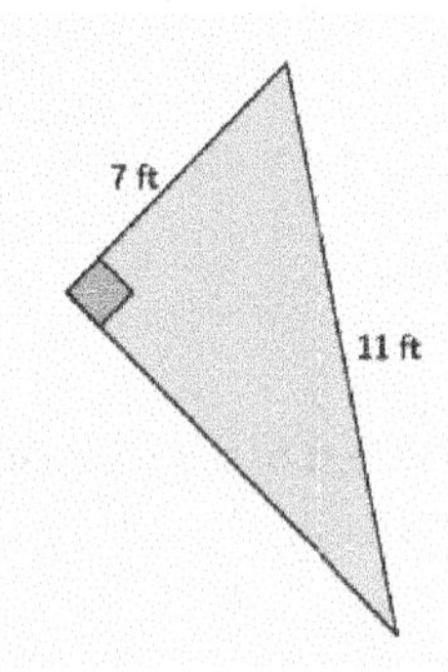

3. Is the triangle with leg lengths of $\sqrt{3}$ cm and 9 cm and hypotenuse of length $\sqrt{84}$ cm a right triangle? Show your work, and answer in a complete sentence.

4. Is the triangle with leg lengths of $\sqrt{7}$ km and 5 km and hypotenuse of length $\sqrt{48}$ km a right triangle? Show your work, and answer in a complete sentence.

5. What is the length of the unknown side of the right triangle shown below? Show your work, and answer in a complete sentence. Provide an exact answer and an approximate answer rounded to the tenths place.

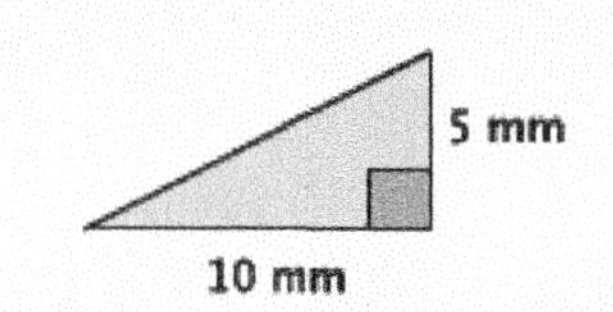

6. Is the triangle with leg lengths of 3 and 6 and hypotenuse of length $\sqrt{45}$ a right triangle? Show your work, and answer in a complete sentence.

7. What is the length of the unknown side of the right triangle shown below? Show your work, and answer in a complete sentence. Provide an exact answer and an approximate answer rounded to the tenths place.

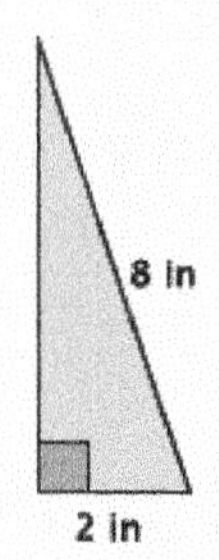

8. Is the triangle with leg lengths of 1 and $\sqrt{3}$ and hypotenuse of length 2 a right triangle? Show your work, and answer in a complete sentence.

9. Corey found the hypotenuse of a right triangle with leg lengths of 2 and 3 to be $\sqrt{13}$. Corey claims that since $\sqrt{13} = 3.61$ when estimating to two decimal digits, that a triangle with leg lengths of 2 and 3 and a hypotenuse of 3.61 is a right triangle. Is he correct? Explain.

10. Explain a proof of the Pythagorean theorem.

11. Explain a proof of the converse of the Pythagorean theorem.

This page intentionally left blank

Lesson 17: Distance on the Coordinate Plane

Classwork

Example 1

What is the distance between the two points A and B on the coordinate plane?

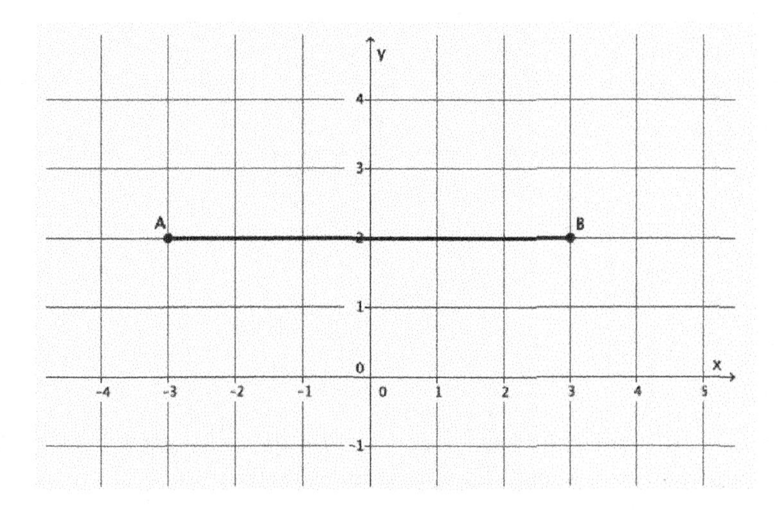

What is the distance between the two points A and B on the coordinate plane?

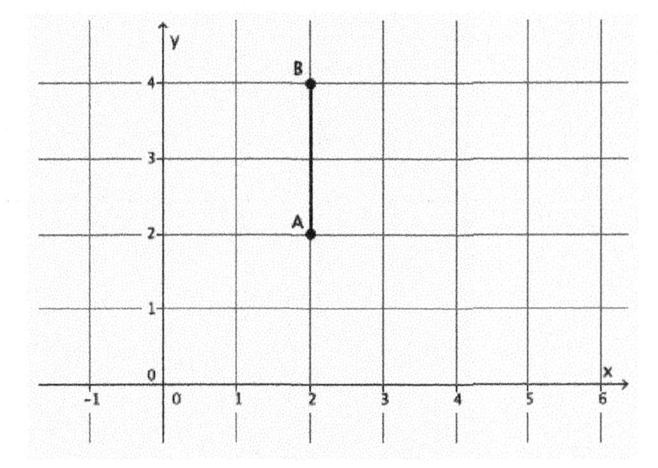

What is the distance between the two points A and B on the coordinate plane? Round your answer to the tenths place.

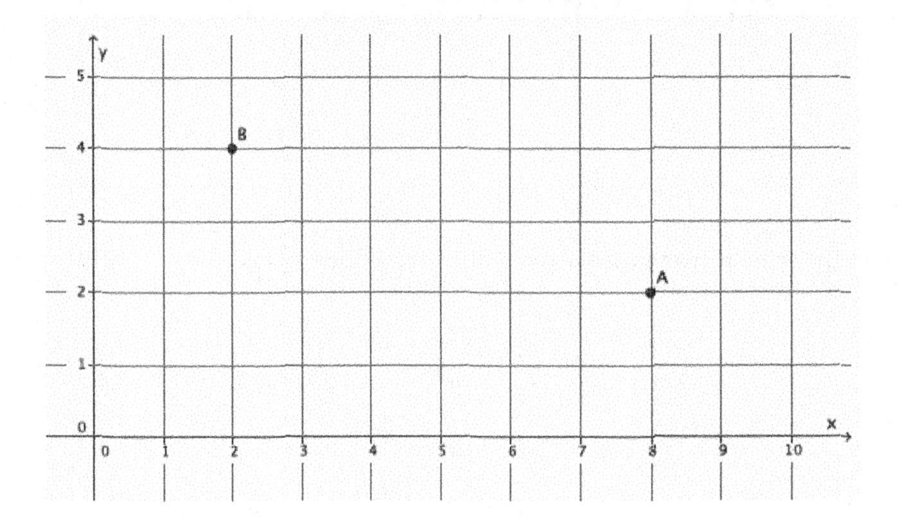

Example 2

Given two points A and B on the coordinate plane, determine the distance between them. First, make an estimate; then, try to find a more precise answer. Round your answer to the tenths place.

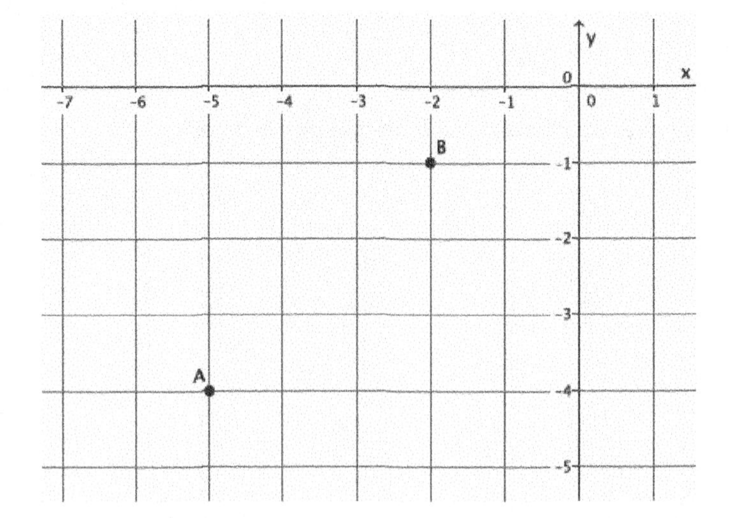

Exercises 1–4

For each of the Exercises 1–4, determine the distance between points A and B on the coordinate plane. Round your answer to the tenths place.

1.

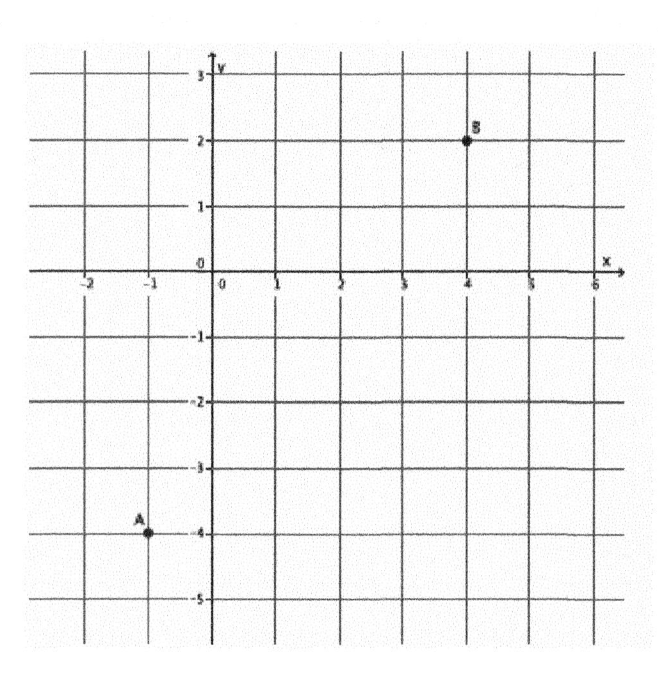

2.

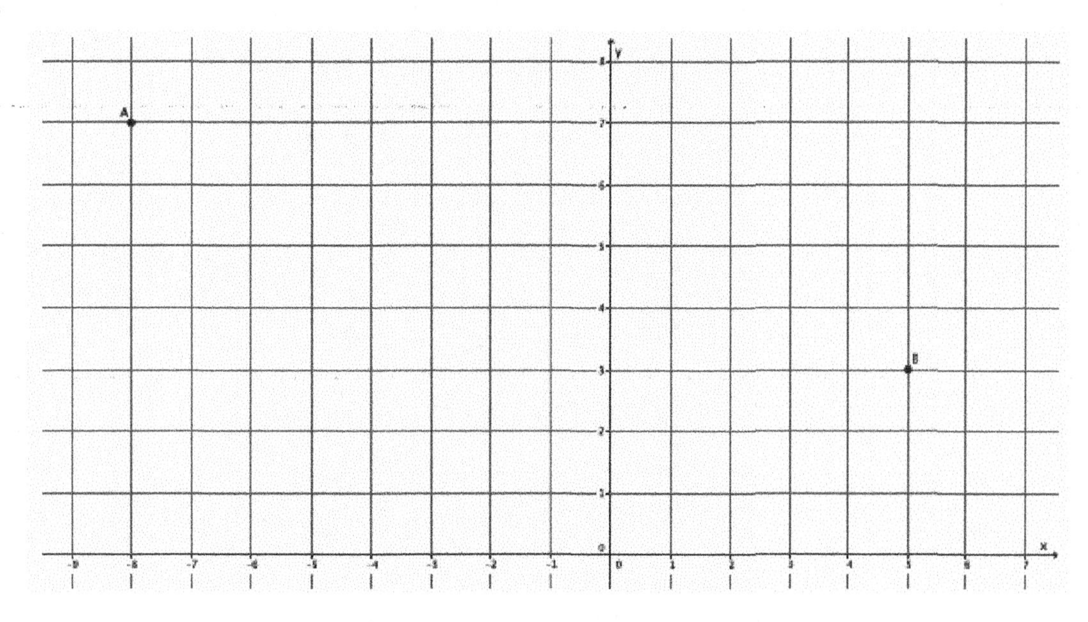

3.

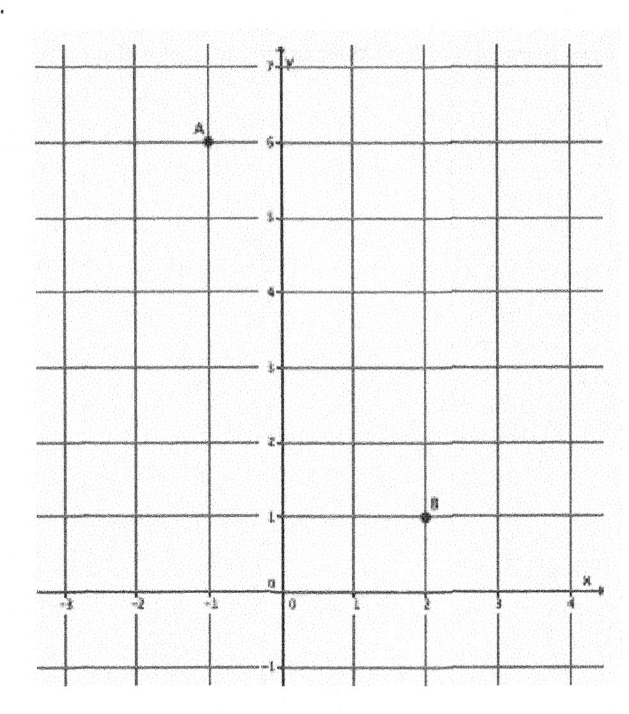

4.

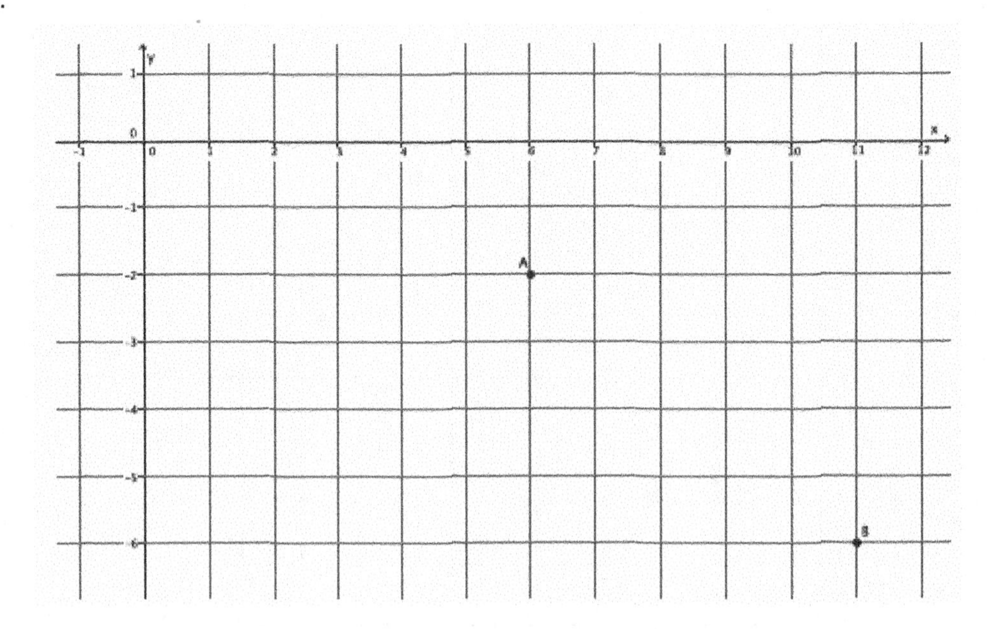

Example 3

Is the triangle formed by the points A, B, C a right triangle?

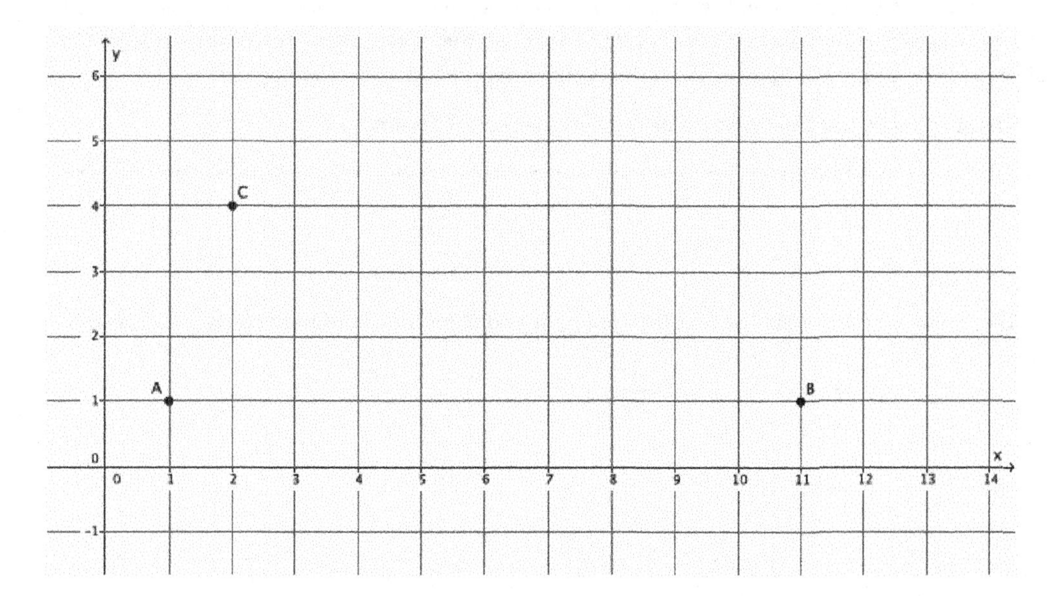

Lesson Summary

To determine the distance between two points on the coordinate plane, begin by connecting the two points. Then, draw a vertical line through one of the points and a horizontal line through the other point. The intersection of the vertical and horizontal lines forms a right triangle to which the Pythagorean theorem can be applied.

To verify if a triangle is a right triangle, use the converse of the Pythagorean theorem.

Problem Set

For each of the Problems 1–4, determine the distance between points A and B on the coordinate plane. Round your answer to the tenths place.

1.

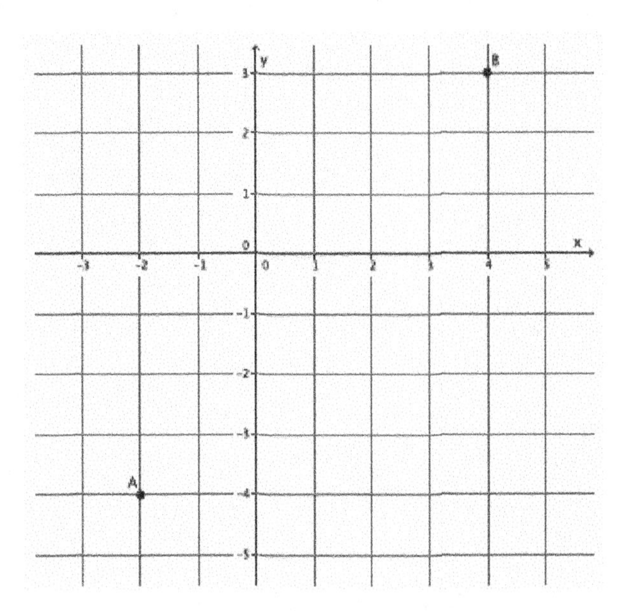

2.

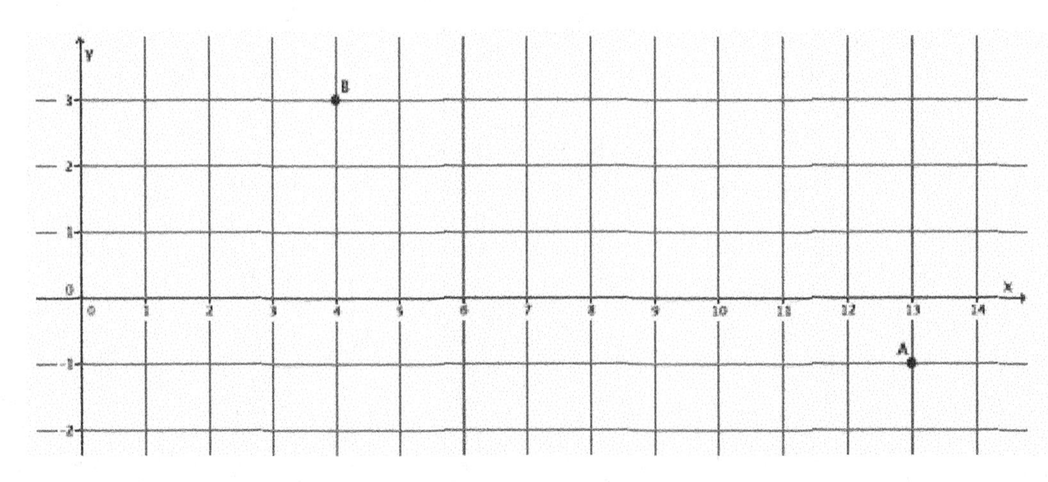

3.

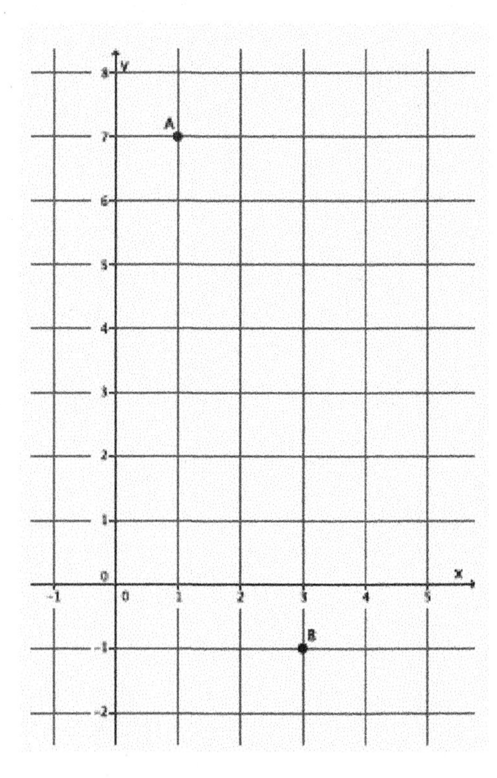

4.

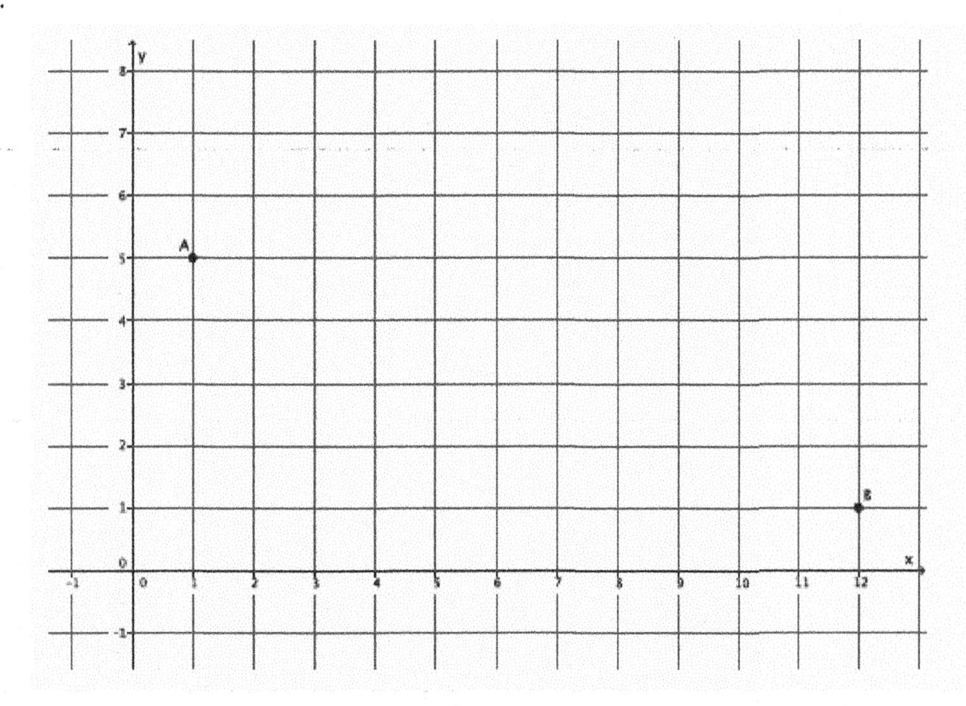

5. Is the triangle formed by points A, B, C a right triangle?

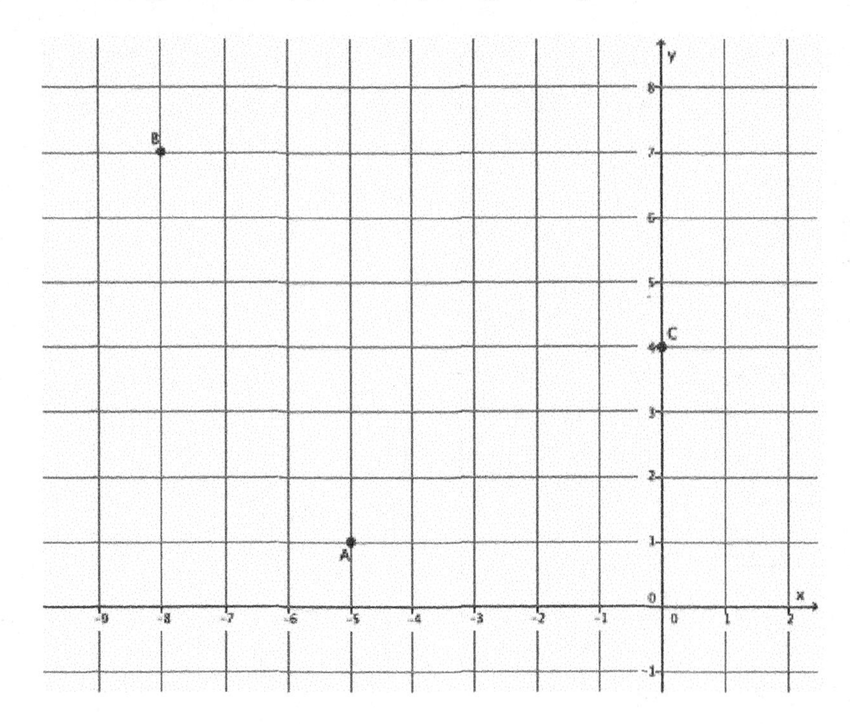

Lesson 18: Applications of the Pythagorean Theorem

Classwork

Exercises

1. The area of the right triangle shown below is 26.46 in^2. What is the perimeter of the right triangle? Round your answer to the tenths place.

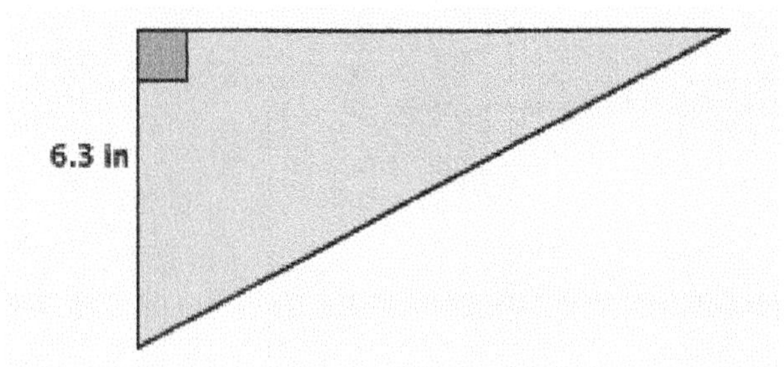

2. The diagram below is a representation of a soccer goal.

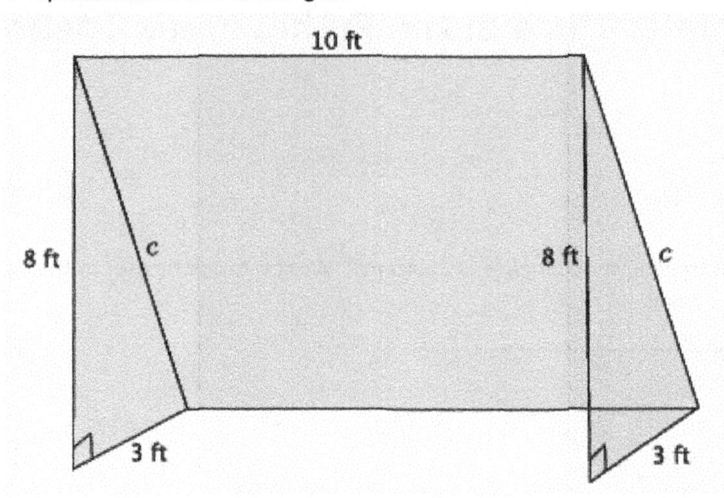

a. Determine the length of the bar, c, that would be needed to provide structure to the goal. Round your answer to the tenths place.

b. How much netting (in square feet) is needed to cover the entire goal?

3. The typical ratio of length to width that is used to produce televisions is $4:3$.

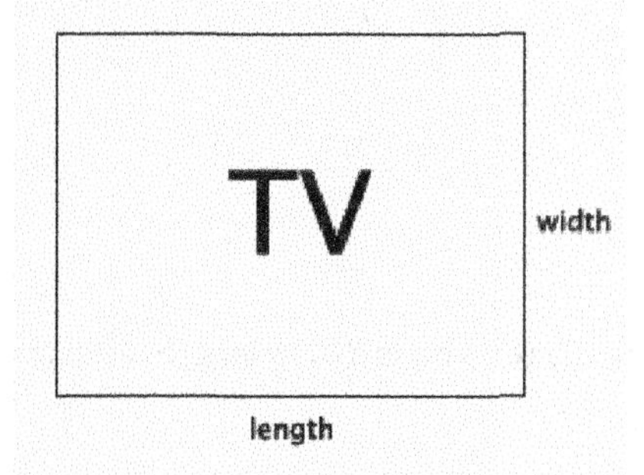

A TV with length 20 inches and width 15 inches, for example, has sides in a $4:3$ ratio; as does any TV with length $4x$ inches and width $3x$ inches for any number x.

a. What is the advertised size of a TV with length 20 inches and width 15 inches?

b. A 42" TV was just given to your family. What are the length and width measurements of the TV?

c. Check that the dimensions you got in part (b) are correct using the Pythagorean theorem.

d. The table that your TV currently rests on is $30''$ in length. Will the new TV fit on the table? Explain.

4. Determine the distance between the following pairs of points. Round your answer to the tenths place. Use graph paper if necessary.

   a. $(7, 4)$ and $(-3, -2)$

   b. $(-5, 2)$ and $(3, 6)$

c. Challenge: (x_1, y_1) and (x_2, y_2). Explain your answer.

5. What length of ladder is needed to reach a height of 7 feet along the wall when the base of the ladder is 4 feet from the wall? Round your answer to the tenths place.

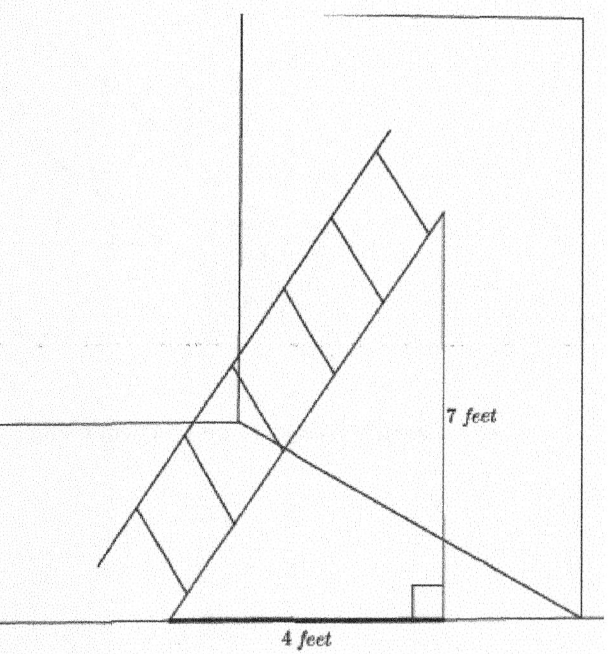

Problem Set

1. A 70" TV is advertised on sale at a local store. What are the length and width of the television?

2. There are two paths that one can use to go from Sarah's house to James' house. One way is to take C Street, and the other way requires you to use A Street and B Street. How much shorter is the direct path along C Street?

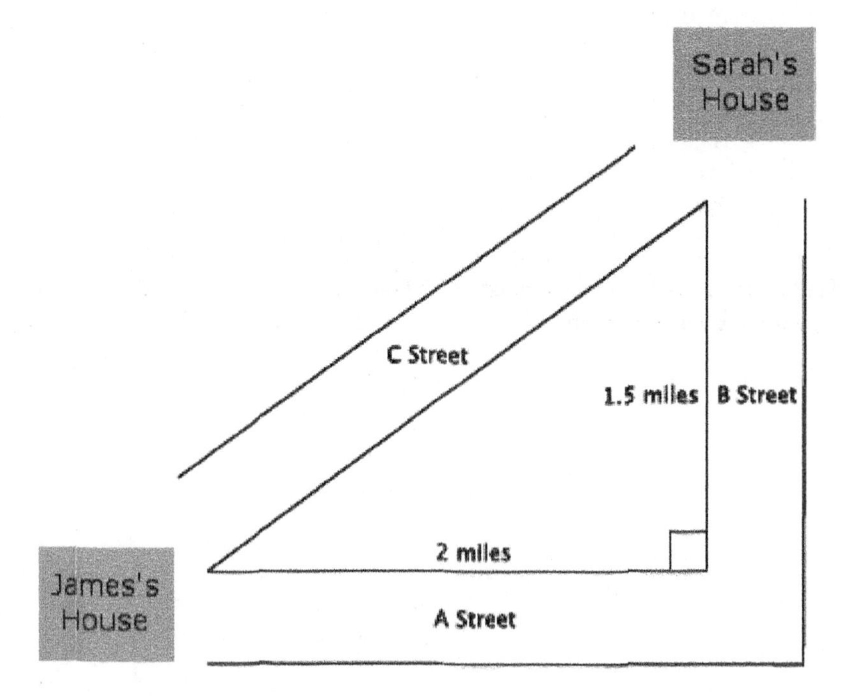

3. An isosceles right triangle refers to a right triangle with equal leg lengths, s, as shown below.

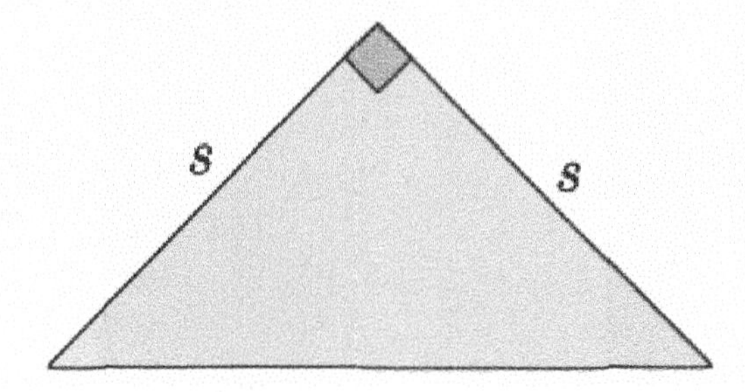

What is the length of the hypotenuse of an isosceles right triangle with a leg length of 9 cm? Write an exact answer using a square root and an approximate answer rounded to the tenths place.

4. The area of the right triangle shown to the right is 66.5 cm^2.
   a. What is the height of the triangle?
   b. What is the perimeter of the right triangle? Round your answer to the tenths place.

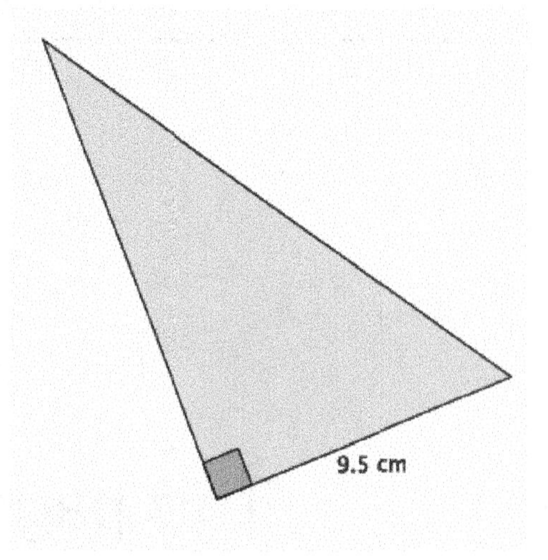

5. What is the distance between points $(1, 9)$ and $(-4, -1)$? Round your answer to the tenths place.

6. An equilateral triangle is shown below. Determine the area of the triangle. Round your answer to the tenths place.

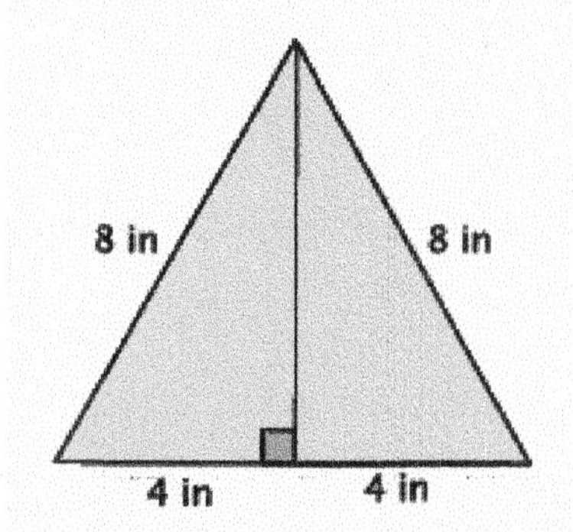

This page intentionally left blank

Lesson 19: Cones and Spheres

Classwork

Exercises 1–2

Note: Figures not drawn to scale.

1. Determine the volume for each figure below.

   a. Write an expression that shows volume in terms of the area of the base, B, and the height of the figure. Explain the meaning of the expression, and then use it to determine the volume of the figure.

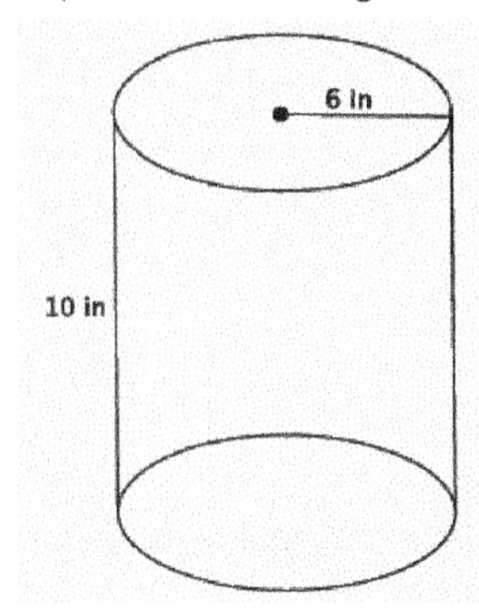

   b. Write an expression that shows volume in terms of the area of the base, B, and the height of the figure. Explain the meaning of the expression, and then use it to determine the volume of the figure.

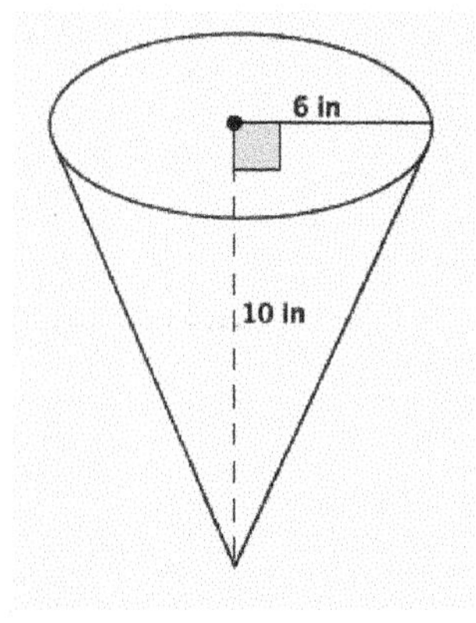

2.

a. Write an expression that shows volume in terms of the area of the base, B, and the height of the figure. Explain the meaning of the expression, and then use it to determine the volume of the figure.

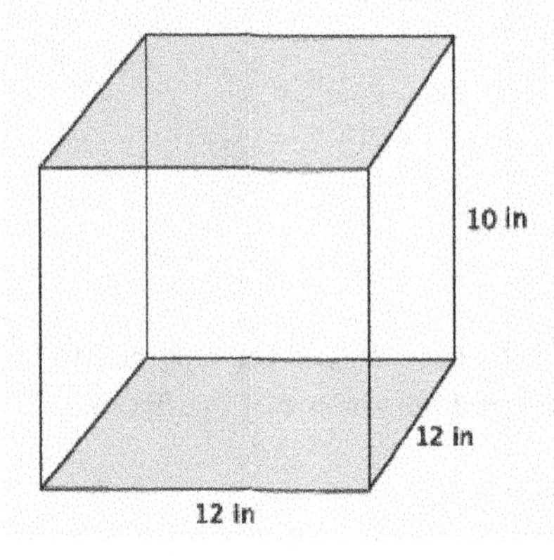

b. The volume of the square pyramid shown below is 480 in^3. What might be a reasonable guess for the formula for the volume of a pyramid? What makes you suggest your particular guess?

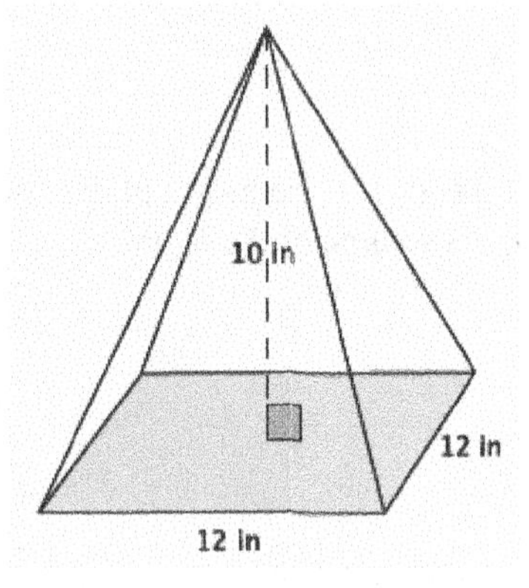

Example 1

State as many facts as you can about a cone.

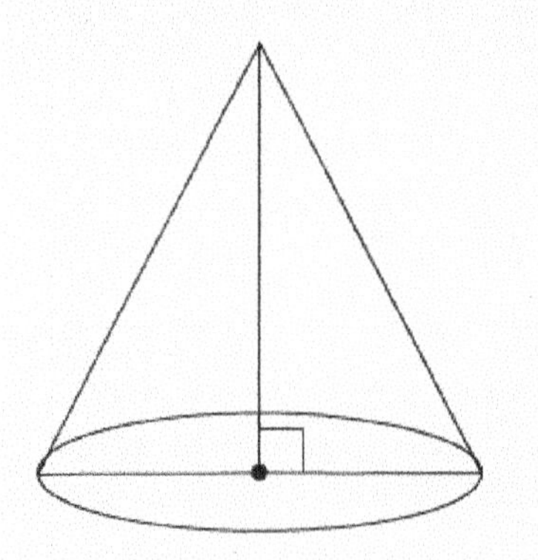

Exercises 3–10

3. What is the lateral length (slant height) of the cone shown below?

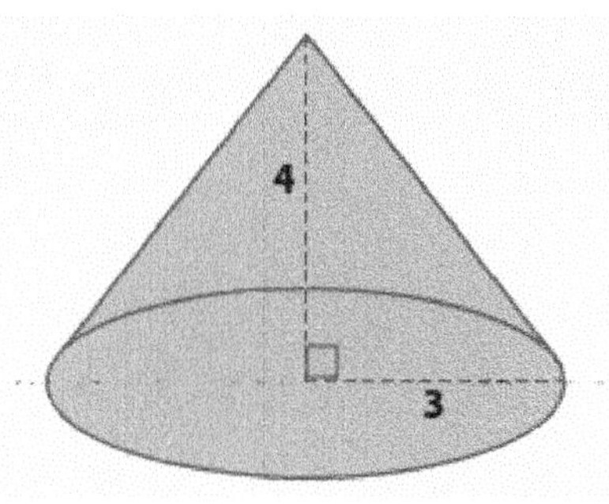

4. Determine the exact volume of the cone shown below.

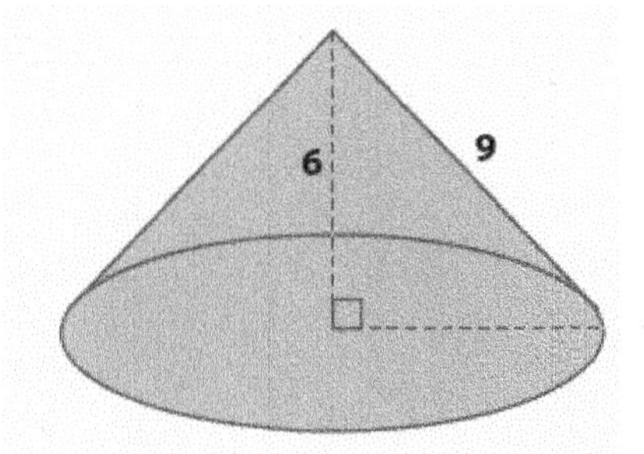

5. What is the lateral length (slant height) of the pyramid shown below? Give an exact square root answer and an approximate answer rounded to the tenths place.

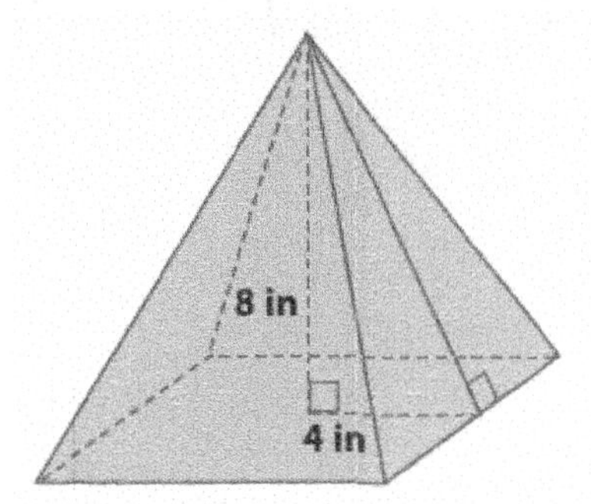

6. Determine the volume of the square pyramid shown below. Give an exact answer using a square root.

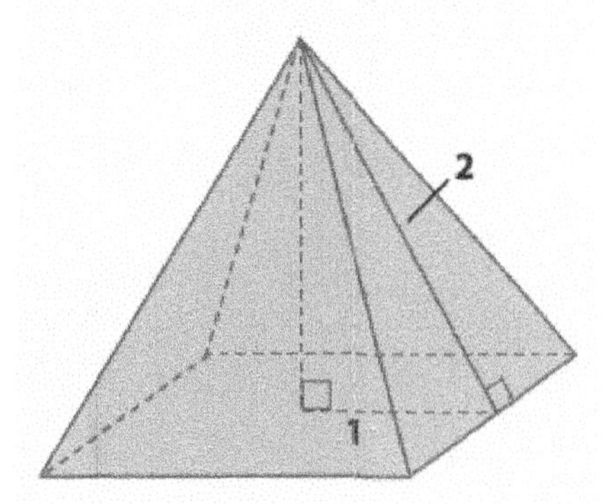

7. What is the length of the chord of the sphere shown below? Give an exact answer using a square root.

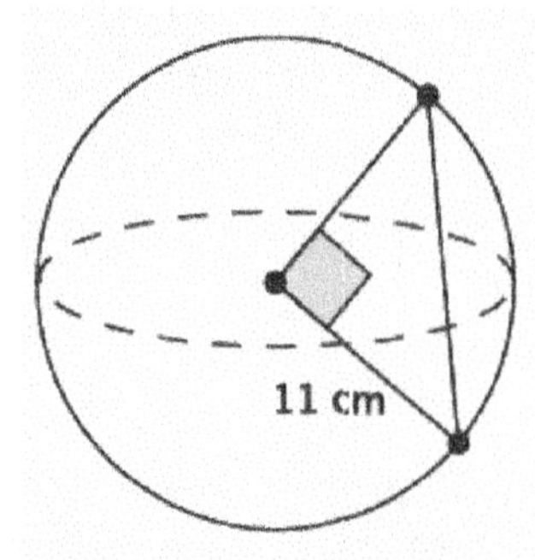

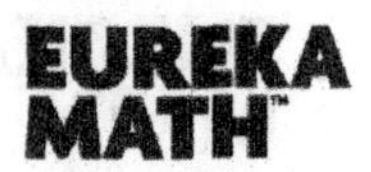

8. What is the length of the chord of the sphere shown below? Give an exact answer using a square root.

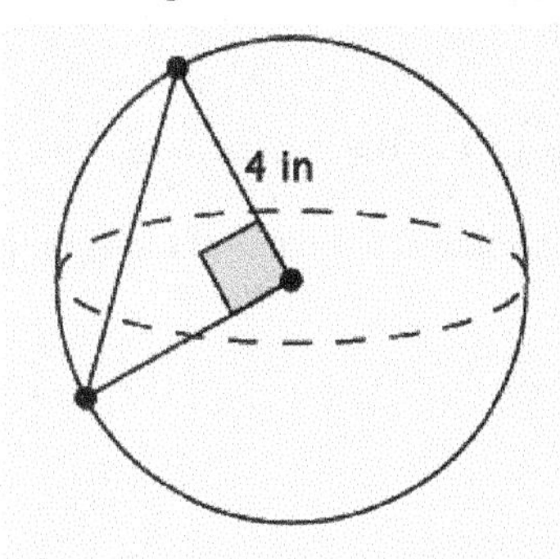

9. What is the volume of the sphere shown below? Give an exact answer using a square root.

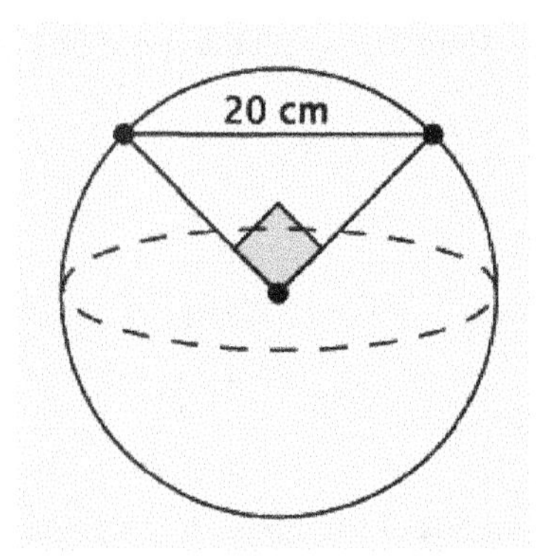

10. What is the volume of the sphere shown below? Give an exact answer using a square root.

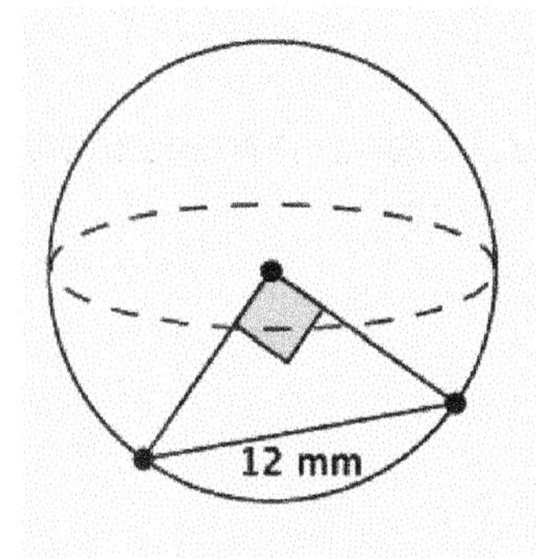

Lesson Summary

The volume formula for a right square pyramid is $V = \frac{1}{3}Bh$, where B is the area of the square base.

The lateral length of a cone, sometimes referred to as the slant height, is the side s, shown in the diagram below.

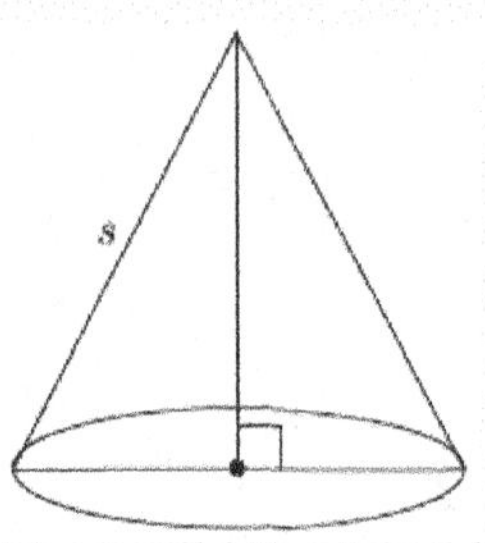

Given the lateral length and the length of the radius, the Pythagorean theorem can be used to determine the height of the cone.

Let O be the center of a circle, and let P and Q be two points on the circle. Then $\overline{PQ}$ is called a chord of the circle.

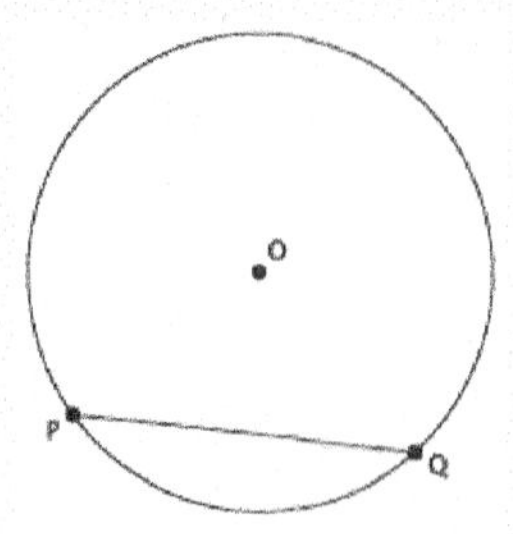

The segments OP and OQ are equal in length because both represent the radius of the circle. If the angle formed by POQ is a right angle, then the Pythagorean theorem can be used to determine the length of the radius when given the length of the chord, or the length of the chord can be determined if given the length of the radius.

Problem Set

1. What is the lateral length (slant height) of the cone shown below? Give an approximate answer rounded to the tenths place.

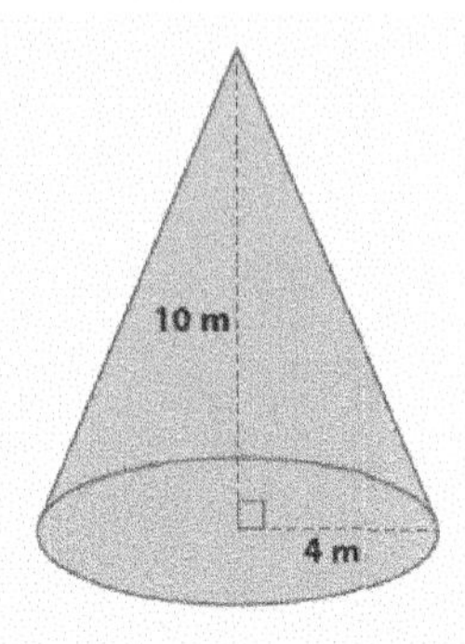

2. What is the volume of the cone shown below? Give an exact answer.

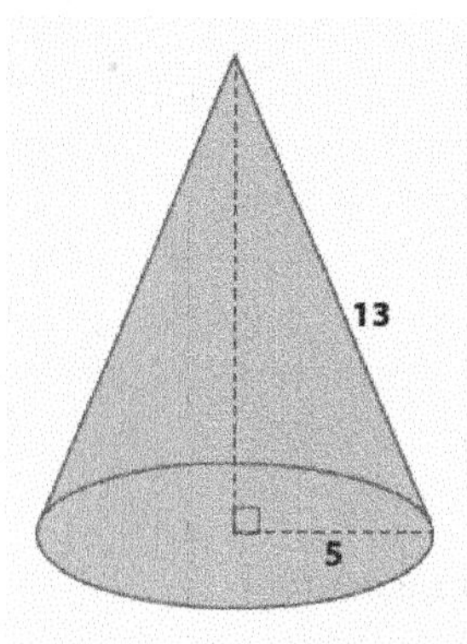

3. Determine the volume and surface area of the square pyramid shown below. Give exact answers.

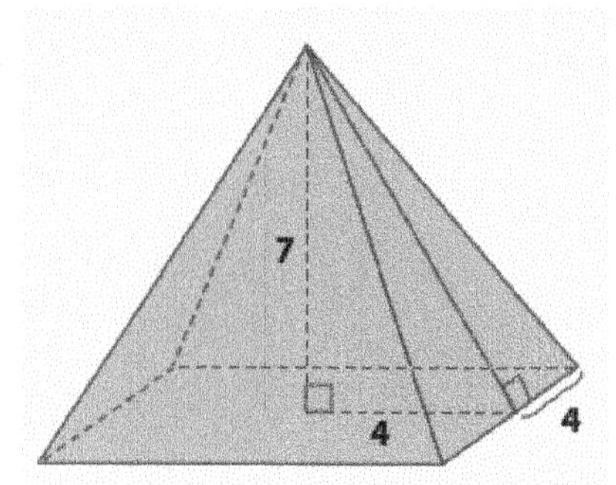

4. Alejandra computed the volume of the cone shown below as $64\pi \text{ cm}^3$. Her work is shown below. Is she correct? If not, explain what she did wrong, and calculate the correct volume of the cone. Give an exact answer.

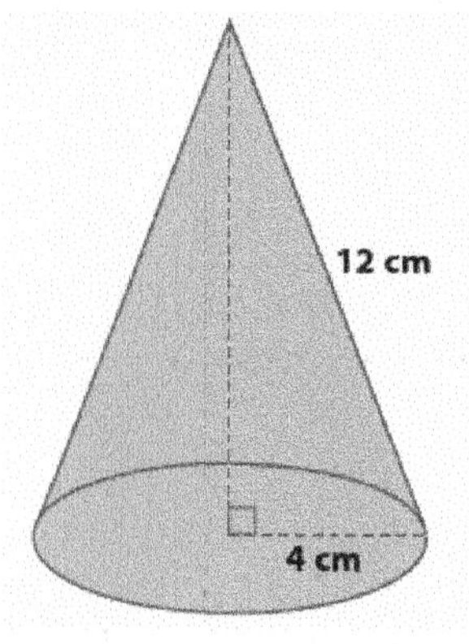

$$V = \frac{1}{3}\pi(4)^2(12)$$
$$= \frac{(16)(12)\pi}{3}$$
$$= 64\pi$$

The volume of the cone is $64\pi \text{ cm}^3$.

5. What is the length of the chord of the sphere shown below? Give an exact answer using a square root.

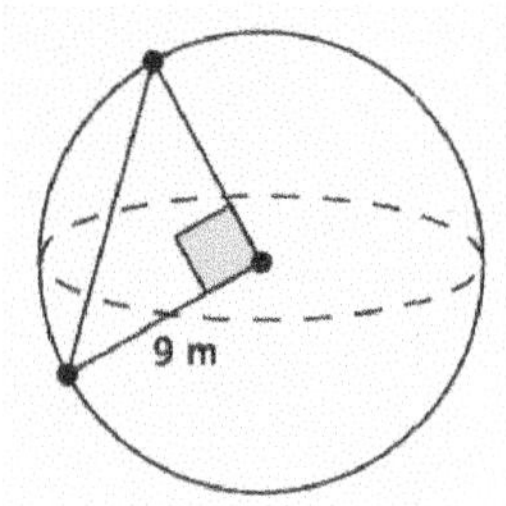

6. What is the volume of the sphere shown below? Give an exact answer using a square root.

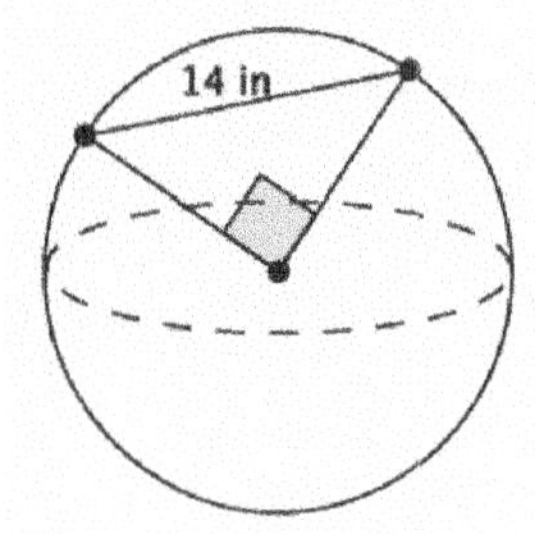

Lesson 20: Truncated Cones

Classwork

Opening Exercise

Examine the bucket below. It has a height of 9 inches and a radius at the top of the bucket of 4 inches.

a. Describe the shape of the bucket. What is it similar to?

b. Estimate the volume of the bucket.

Example 1

Determine the volume of the truncated cone shown below.

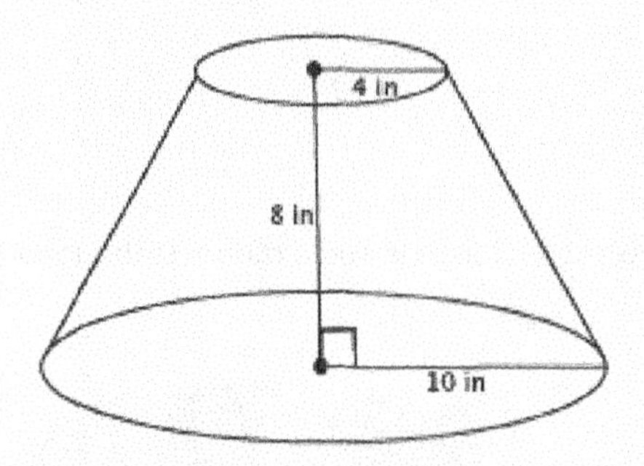

Exercises 1–5

1. Find the volume of the truncated cone.

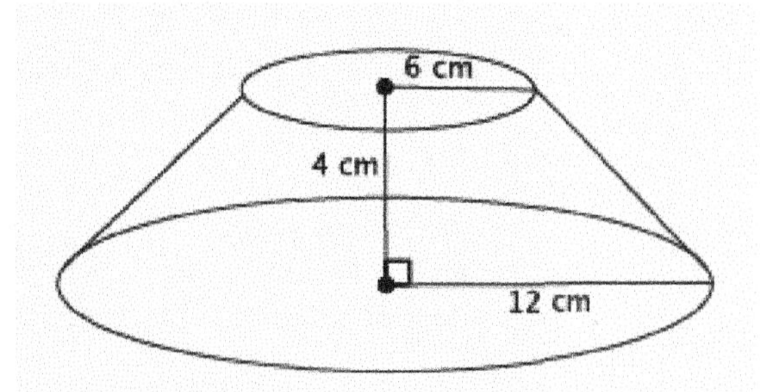

   a. Write a proportion that will allow you to determine the height of the cone that has been removed. Explain what all parts of the proportion represent.

   b. Solve your proportion to determine the height of the cone that has been removed.

   c. Write an expression that can be used to determine the volume of the truncated cone. Explain what each part of the expression represents.

   d. Calculate the volume of the truncated cone.

2. Find the volume of the truncated cone.

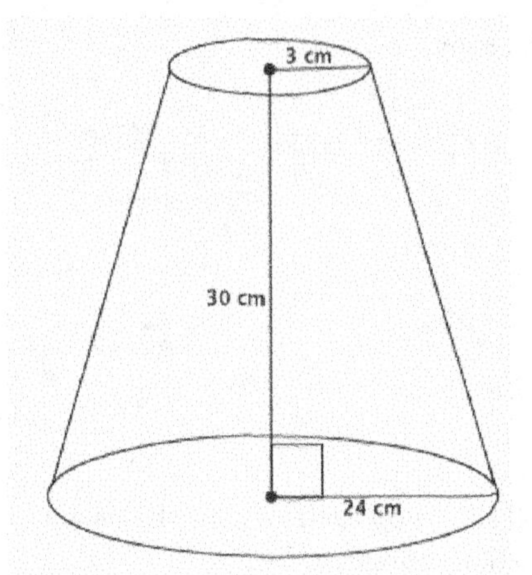

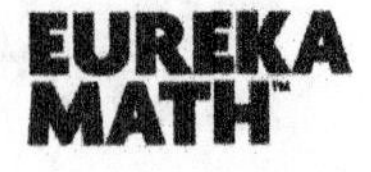

3. Find the volume of the truncated pyramid with a square base.

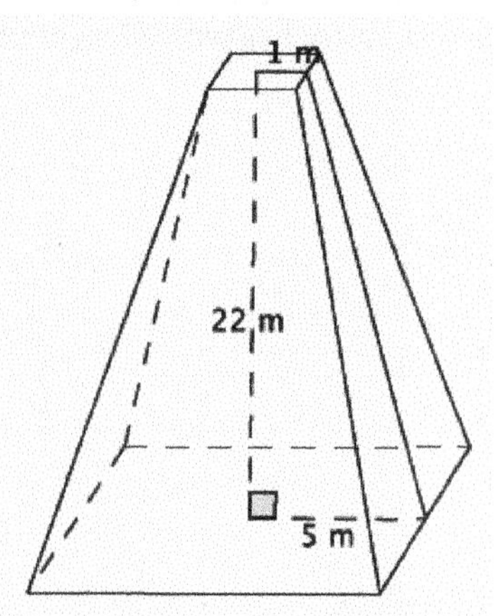

a. Write a proportion that will allow you to determine the height of the cone that has been removed. Explain what all parts of the proportion represent.

b. Solve your proportion to determine the height of the pyramid that has been removed.

c. Write an expression that can be used to determine the volume of the truncated pyramid. Explain what each part of the expression represents.

d. Calculate the volume of the truncated pyramid.

4. A pastry bag is a tool used to decorate cakes and cupcakes. Pastry bags take the form of a truncated cone when filled with icing. What is the volume of a pastry bag with a height of 6 inches, large radius of 2 inches, and small radius of 0.5 inches?

5. Explain in your own words what a truncated cone is and how to determine its volume.

Lesson Summary

A truncated cone or pyramid is the solid obtained by removing the top portion of a cone or a pyramid above a plane parallel to its base. Shown below on the left is a truncated cone. A truncated cone with the top portion still attached is shown below on the right.

Truncated cone:

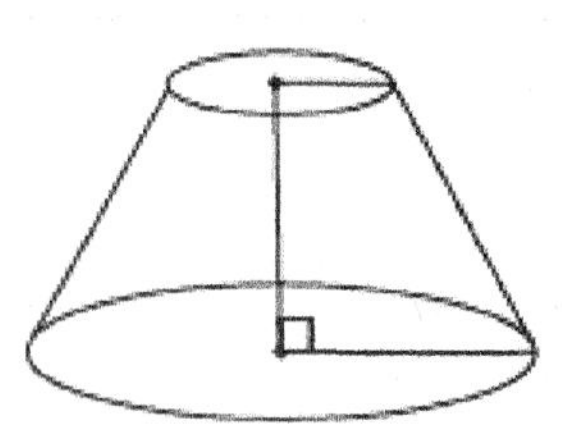

Truncated cone with top portion attached:

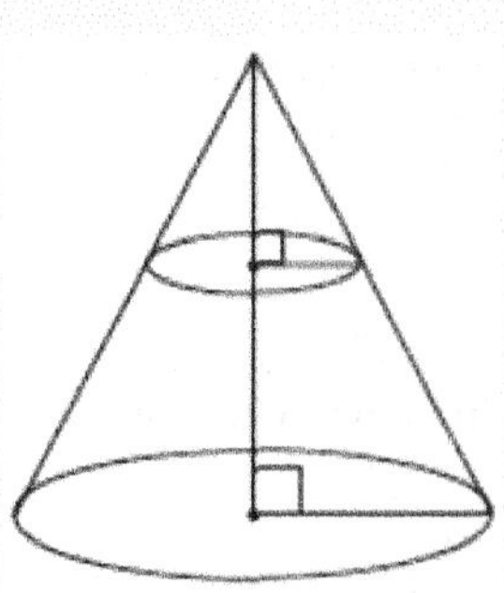

To determine the volume of a truncated cone, you must first determine the height of the portion of the cone that has been removed using ratios that represent the corresponding sides of the right triangles. Next, determine the volume of the portion of the cone that has been removed and the volume of the truncated cone with the top portion attached. Finally, subtract the volume of the cone that represents the portion that has been removed from the complete cone. The difference represents the volume of the truncated cone.

Pictorially,

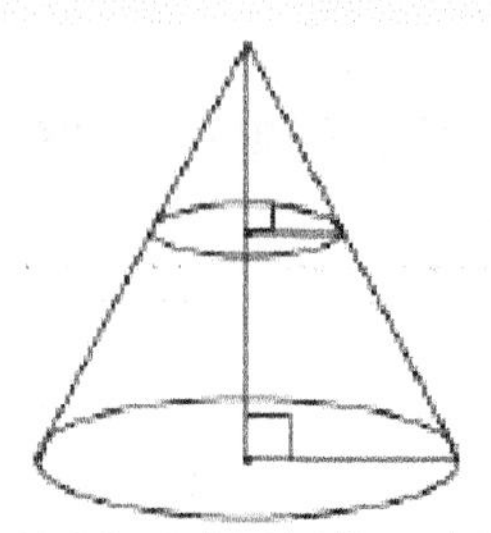

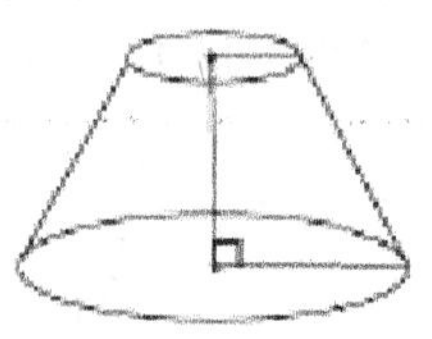

Problem Set

1. Find the volume of the truncated cone.
   a. Write a proportion that will allow you to determine the height of the cone that has been removed. Explain what each part of the proportion represents.
   b. Solve your proportion to determine the height of the cone that has been removed.
   c. Show a fact about the volume of the truncated cone using an expression. Explain what each part of the expression represents.
   d. Calculate the volume of the truncated cone.

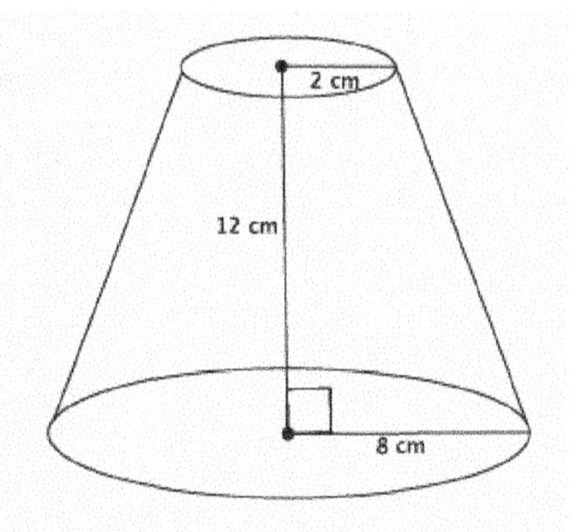

2. Find the volume of the truncated cone.

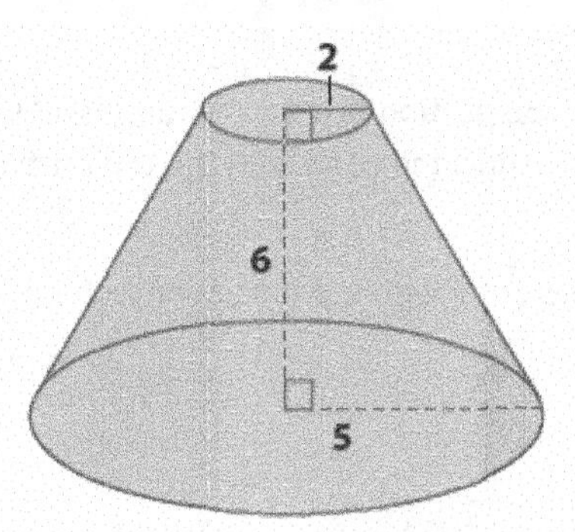

3. Find the volume of the truncated pyramid with a square base.

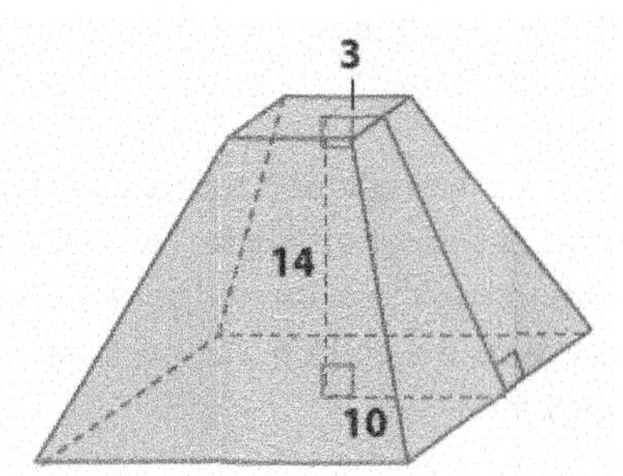

4. Find the volume of the truncated pyramid with a square base. Note: 3 mm is the distance from the center to the edge of the square at the top of the figure.

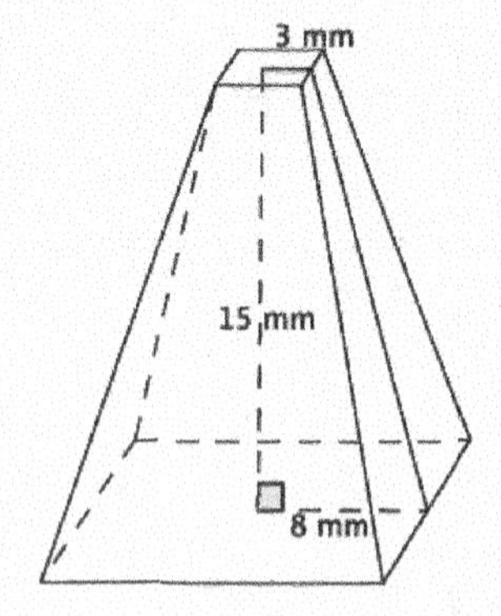

5. Find the volume of the truncated pyramid with a square base. Note: 0.5 cm is the distance from the center to the edge of the square at the top of the figure.

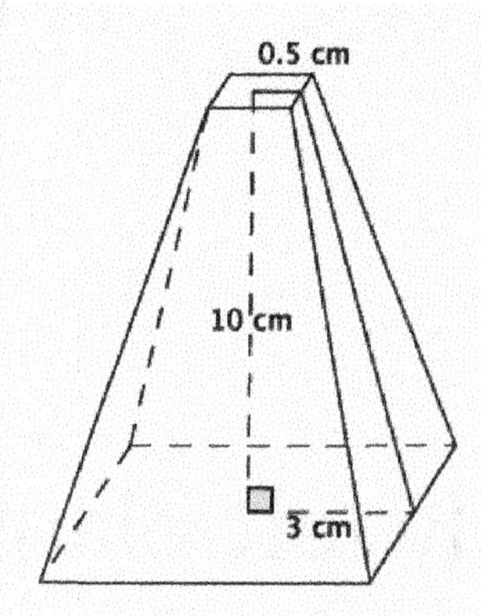

6. Explain how to find the volume of a truncated cone.

7. Challenge: Find the volume of the truncated cone.

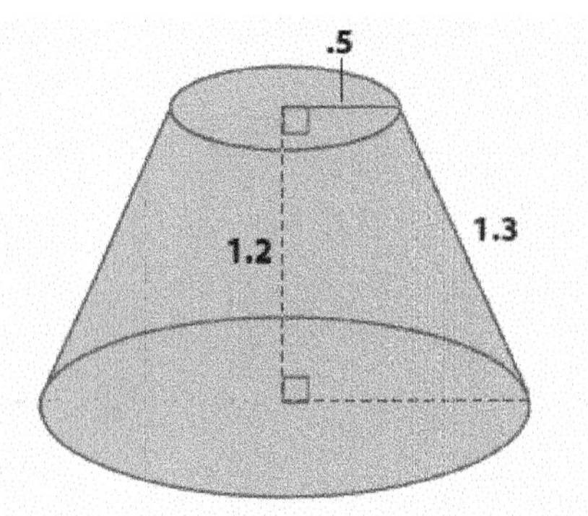

This page intentionally left blank

Lesson 21: Volume of Composite Solids

Classwork

Exercises 1–4

1.

   a. Write an expression that can be used to find the volume of the chest shown below. Explain what each part of your expression represents. (Assume the ends of the top portion of the chest are semicircular.)

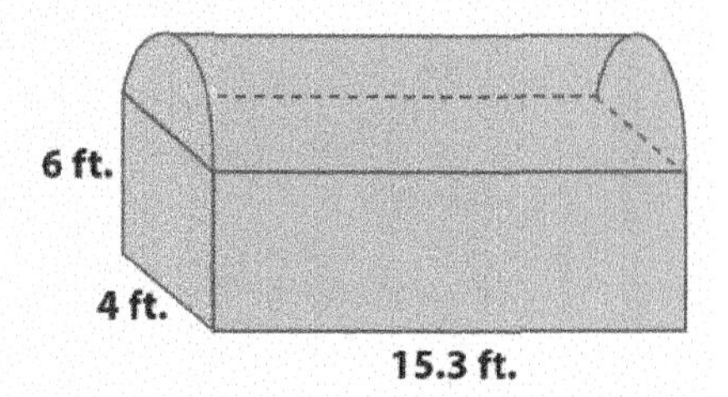

   b. What is the approximate volume of the chest shown above? Use 3.14 for an approximation of π. Round your final answer to the tenths place.

2.

a. Write an expression for finding the volume of the figure, an ice cream cone and scoop, shown below. Explain what each part of your expression represents. (Assume the sphere just touches the base of the cone.)

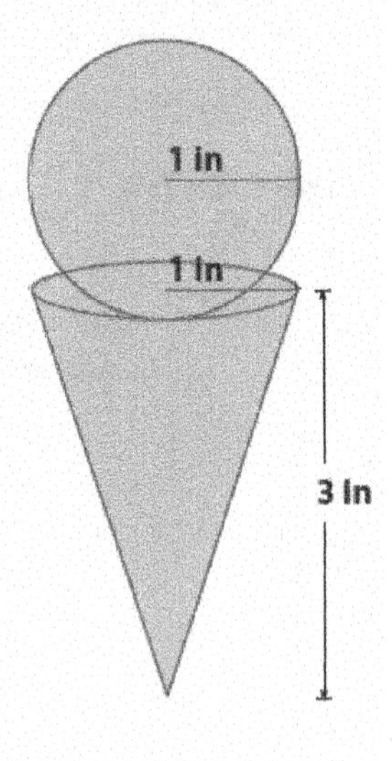

b. Assuming every part of the cone can be filled with ice cream, what is the exact and approximate volume of the cone and scoop? (Recall that exact answers are left in terms of π, and approximate answers use 3.14 for π). Round your approximate answer to the hundredths place.

3.

a. Write an expression for finding the volume of the figure shown below. Explain what each part of your expression represents.

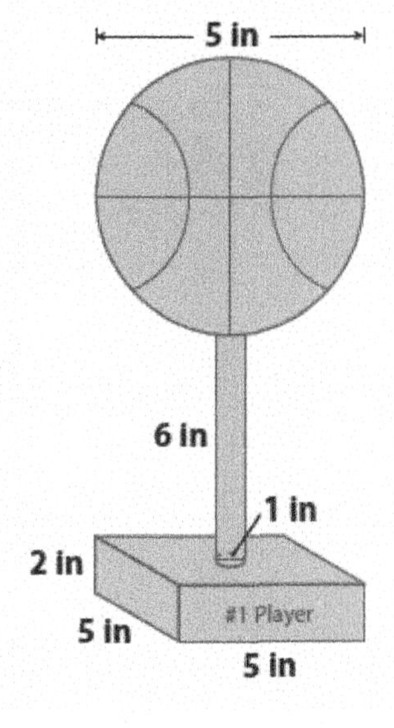

b. Every part of the trophy shown is solid and made out of silver. How much silver is used to produce one trophy? Give an exact and approximate answer rounded to the hundredths place.

4. Use the diagram of scoops below to answer parts (a) and (b).

   a. Order the scoops from least to greatest in terms of their volumes. Each scoop is measured in inches. (Assume the third scoop is hemi-spherical.)

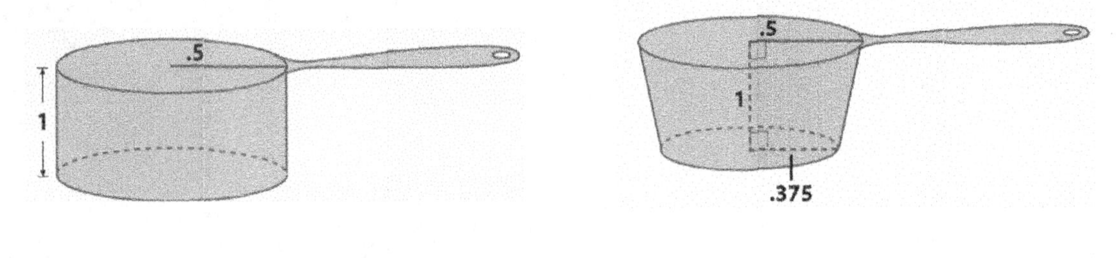

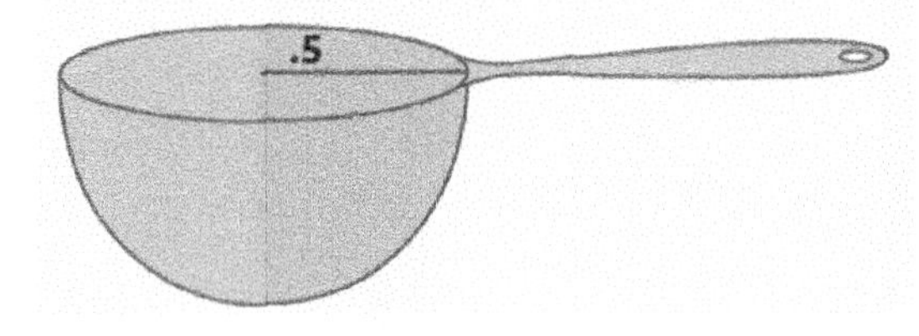

b. How many of each scoop would be needed to add a half-cup of sugar to a cupcake mixture? (One-half cup is approximately 7 in^3.) Round your answer to a whole number of scoops.

Lesson Summary

Composite solids are figures comprising more than one solid. Volumes of composite solids can be added as long as no parts of the solids overlap. That is, they touch only at their boundaries.

Problem Set

1. What volume of sand is required to completely fill up the hourglass shown below? Note: 12 m is the height of the truncated cone, not the lateral length of the cone.

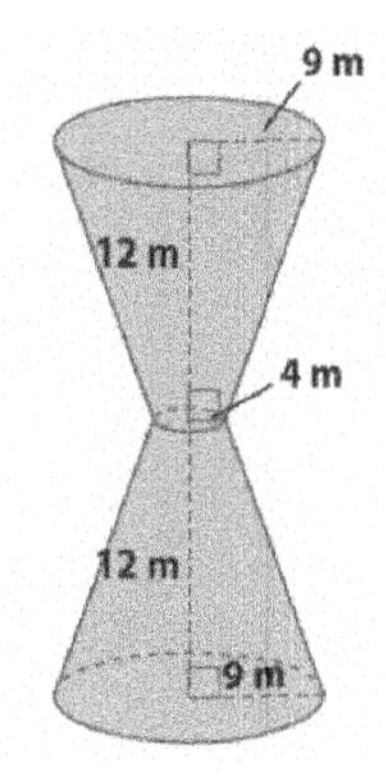

2.
   a. Write an expression for finding the volume of the prism with the pyramid portion removed. Explain what each part of your expression represents.

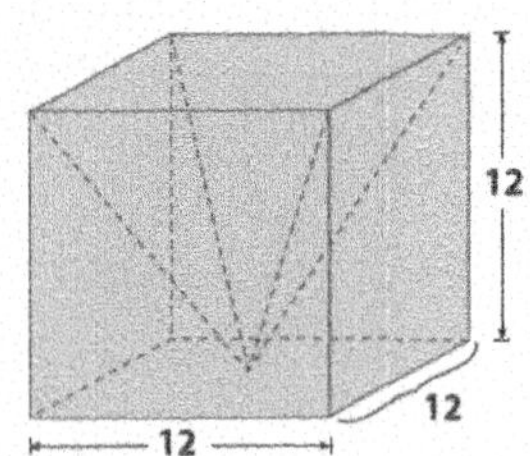

   b. What is the volume of the prism shown above with the pyramid portion removed?

3.
   a. Write an expression for finding the volume of the funnel shown to the right. Explain what each part of your expression represents.
   b. Determine the exact volume of the funnel.

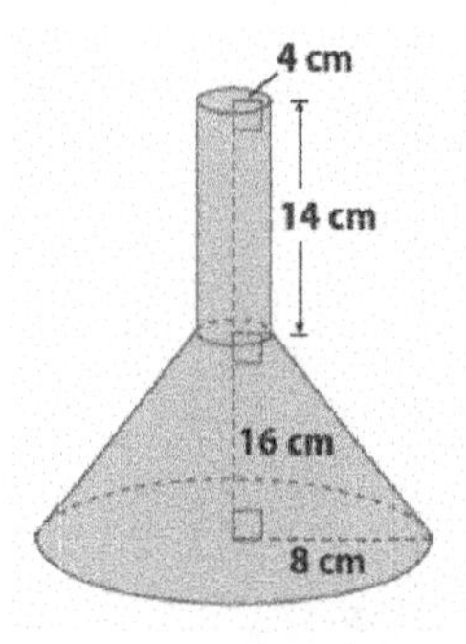

4. What is the approximate volume of the rectangular prism with a cylindrical hole shown below? Use 3.14 for π. Round your answer to the tenths place.

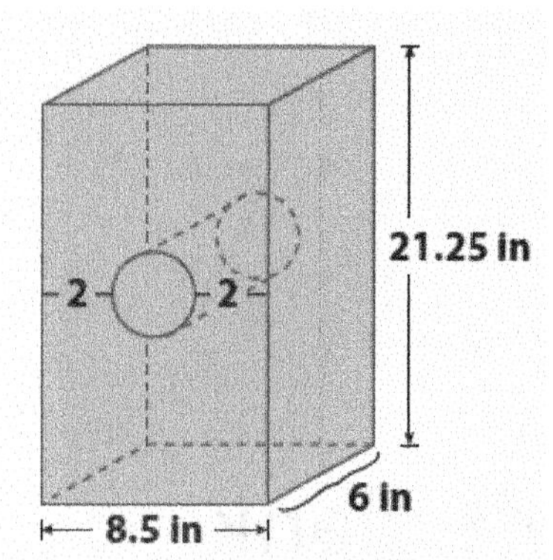

5. A layered cake is being made to celebrate the end of the school year. What is the exact total volume of the cake shown below?

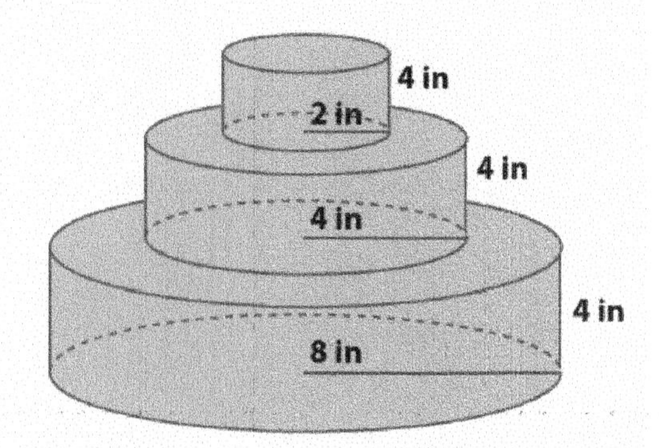

This page intentionally left blank

Lesson 22: Average Rate of Change

Classwork

Exercise

The height of a container in the shape of a circular cone is 7.5 ft., and the radius of its base is 3 ft., as shown. What is the total volume of the cone?

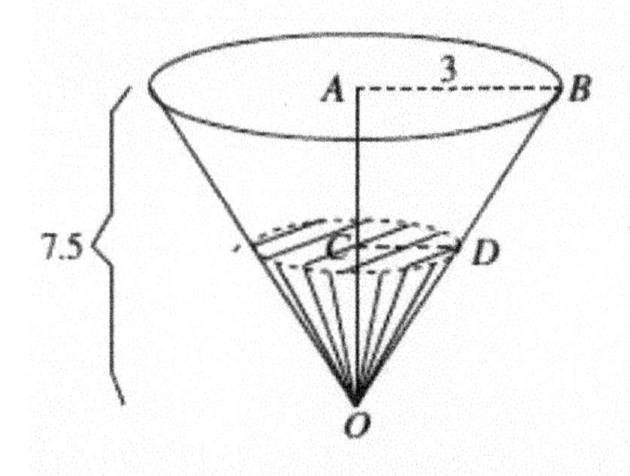

Time (in minutes)	Water Level (in feet)
	1
	2
	3
	4
	5
	6
	7
	7.5

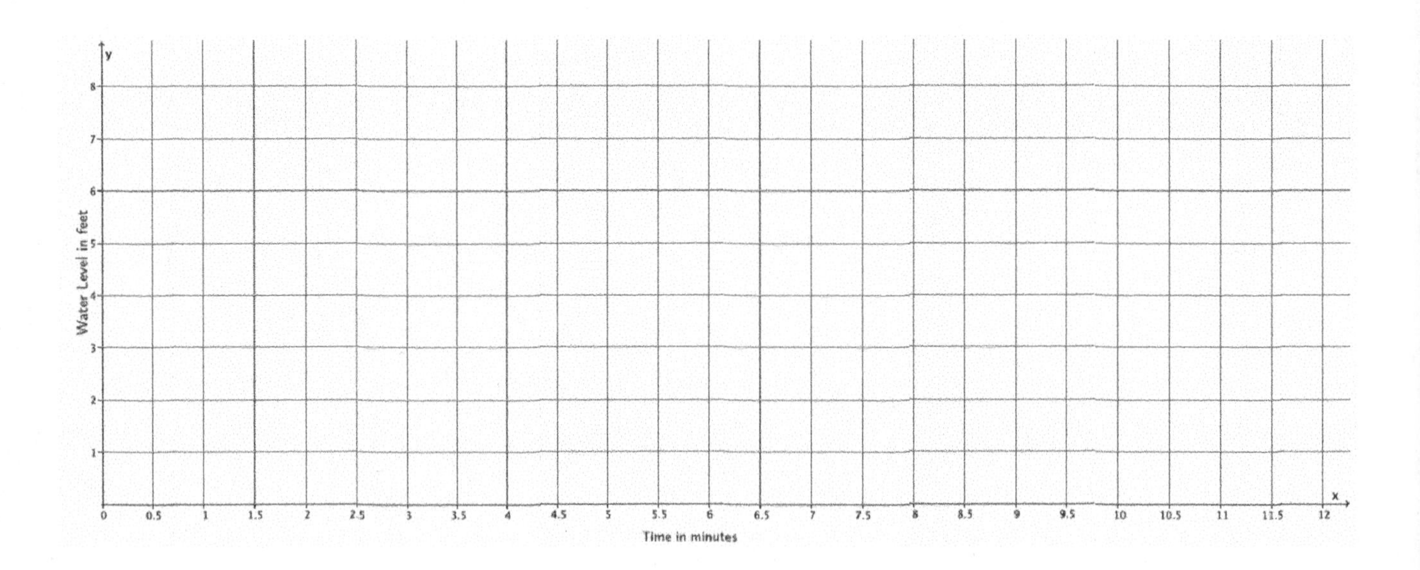

Problem Set

1. Complete the table below for more intervals of water levels of the cone discussed in class. Then, graph the data on a coordinate plane.

Time (in minutes)	Water Level (in feet)
	1
	1.5
	2
	2.5
	3
	3.5
	4
	4.5
	5
	5.5
	6
	6.5
	7
	7.5

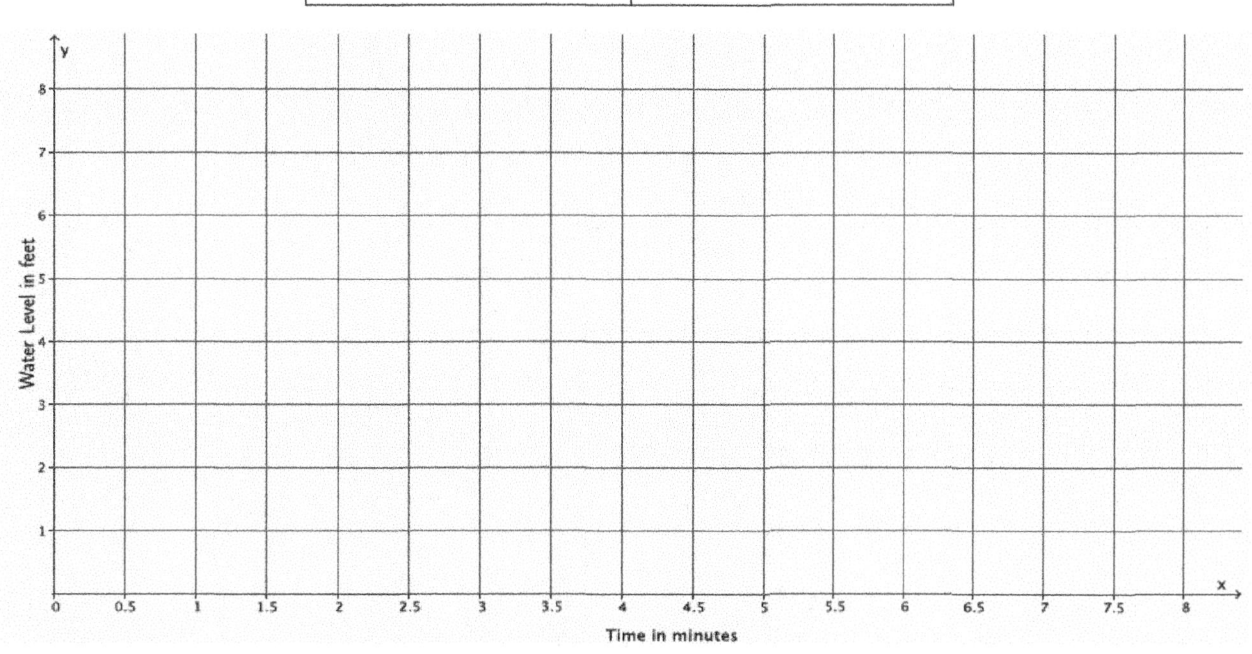

2. Complete the table below, and graph the data on a coordinate plane. Compare the graphs from Problems 1 and 2. What do you notice? If you could write a rule to describe the function of the rate of change of the water level of the cone, what might the rule include?

x	$\sqrt{x}$
1	
4	
9	
16	
25	
36	
49	
64	
81	
100	

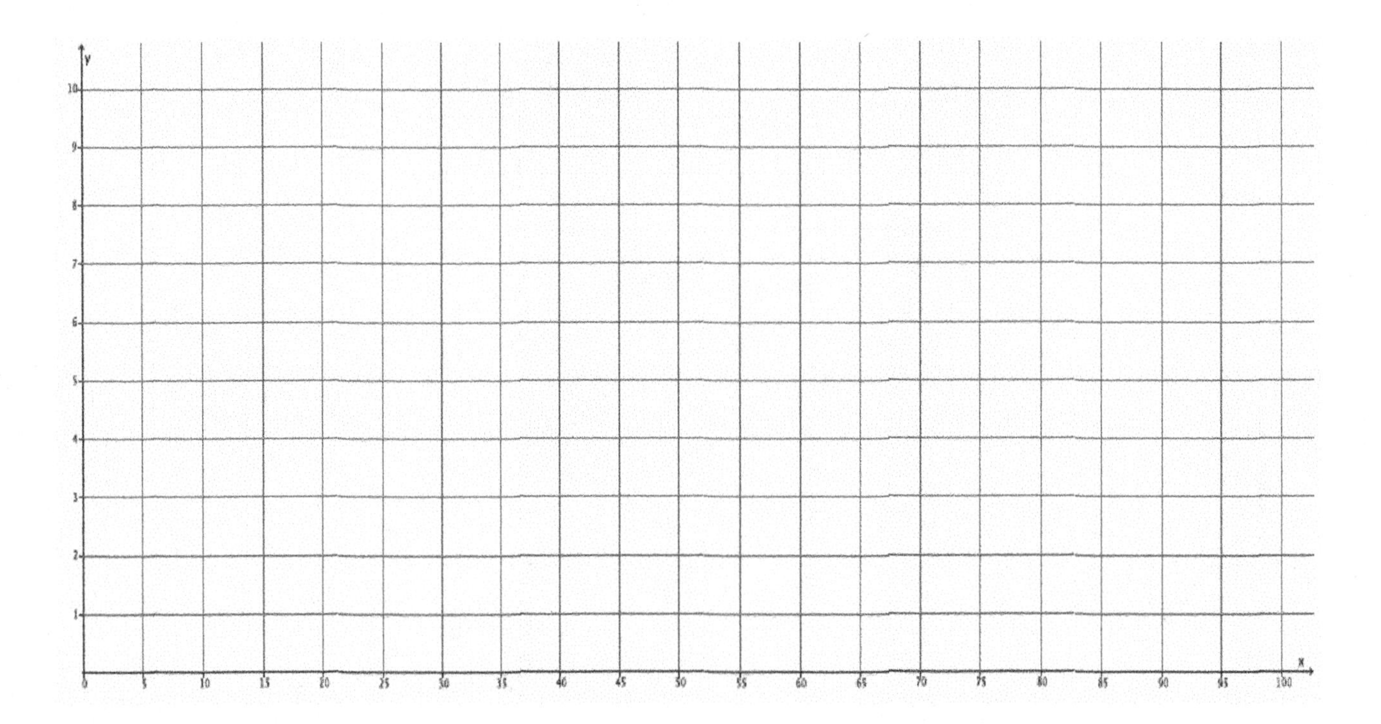

3. Describe, intuitively, the rate of change of the water level if the container being filled were a cylinder. Would we get the same results as with the cone? Why or why not? Sketch a graph of what filling the cylinder might look like, and explain how the graph relates to your answer.

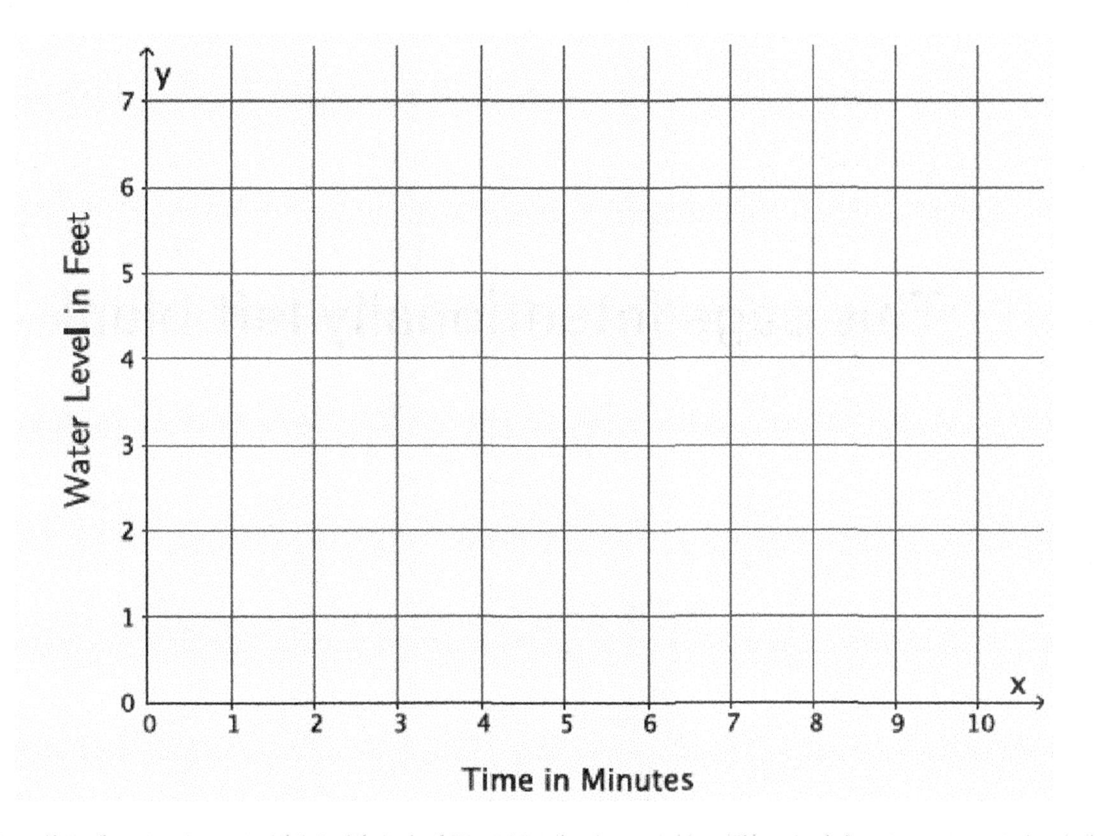

4. Describe, intuitively, the rate of change if the container being filled were a sphere. Would we get the same results as with the cone? Why or why not?

This page intentionally left blank

Lesson 23: Nonlinear Motion

Classwork

Mathematical Modeling Exercise

A ladder of length L ft. leaning against a wall is sliding down. The ladder starts off being flush with (right up against) the wall. The top of the ladder slides down the vertical wall at a constant speed of v ft. per second. Let the ladder in the position L_1 slide down to position L_2 after 1 second, as shown below.

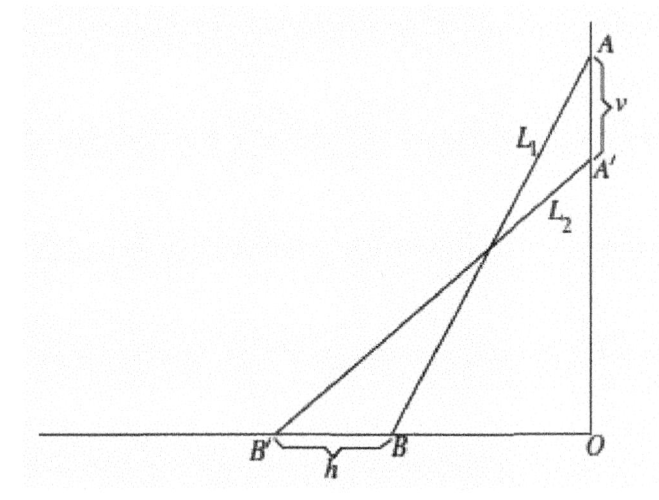

Will the bottom of the ladder move at a constant rate away from point O?

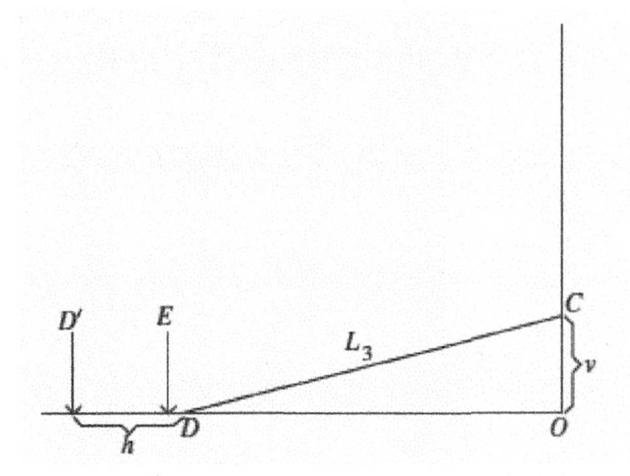

Consider the three right triangles shown below, specifically the change in the length of the base as the height decreases in increments of 1 ft.

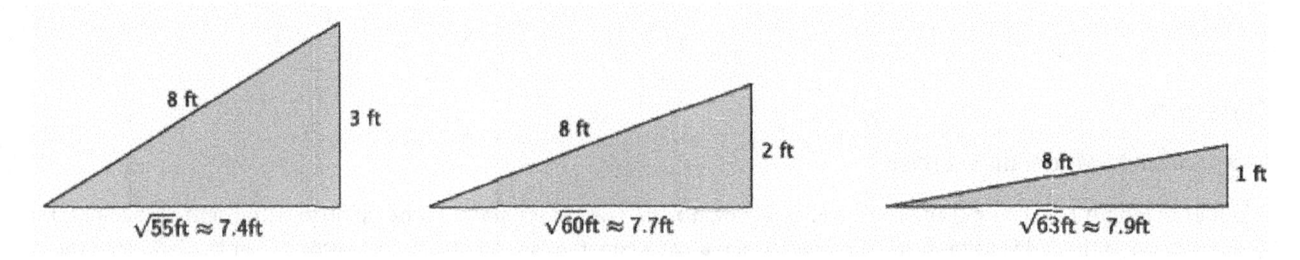

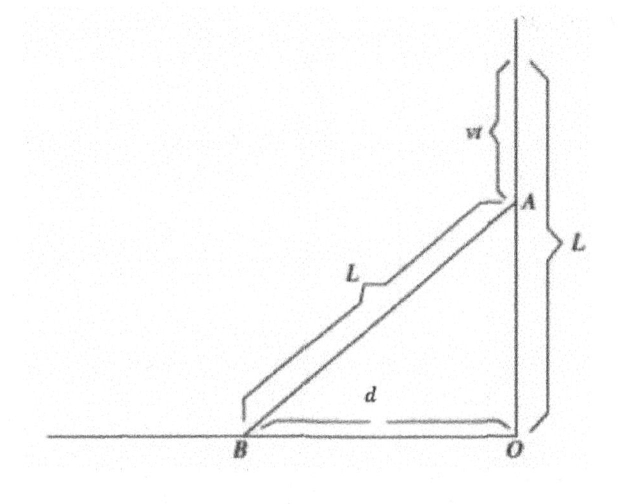

Input (in seconds) t	Output (in feet) $d = \sqrt{225 - (15 - t)^2}$
0	
1	
3	
4	
7	
8	
14	
15	

Problem Set

1. Suppose the ladder is 10 feet long, and the top of the ladder is sliding down the wall at a rate of 0.8 ft. per second. Compute the average rate of change in the position of the bottom of the ladder over the intervals of time from 0 to 0.5 seconds, 3 to 3.5 seconds, 7 to 7.5 seconds, 9.5 to 10 seconds, and 12 to 12.5 seconds. How do you interpret these numbers?

Input (in seconds) t	**Output (in feet)** $d = \sqrt{100 - (10 - 0.8t)^2}$
0	
0.5	
3	
3.5	
7	
7.5	
9.5	
10	
12	
12.5	

2. Will any length of ladder, L, and any constant speed of sliding of the top of the ladder, v ft. per second, ever produce a constant rate of change in the position of the bottom of the ladder? Explain.

This page intentionally left blank